SCIENTIFICA
PRESENTS

PETE RIDISH and MOLLY KEWELL
IN

THE AMAZEMENT PARK

Louise Petheram

Published in 2005 by:
Nelson Thornes Ltd
Delta Place
27 Bath Road
CHELTENHAM
GL53 7TH
United Kingdom

05 06 07 08 09 / 10 9 8 7 6 5 4 3 2 1

A catalogue record for this book is available from the British Library

ISBN 0 7487 9014 4

Cover illustration by Gary Andrews
Illustrations by Oxford Designers and Illustrators
Flick strip illustrations by Darren Watts
Design and page make-up by Darren Watts

Printed and bound in Great Britain by Scotprint

Contents

 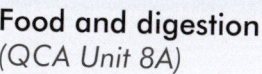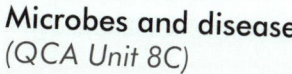

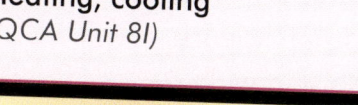

Introduction

The Amazement Park is one of three science readers that you can use with the *Scientifica* books. All the articles in *The Amazement Park* are linked to science topics that you will cover in the *Scientifica* Year 8 book. The Contents page shows you which Unit in the *Scientifica* book each article links to. The articles focus on some of the 'amazing' things we can do with science, including some of the 'amazing' ideas that modern scientists are working on right now!

The articles are ideal to help you with the 'Ideas and Evidence' part of your science course. You will find out how scientists work today, making discoveries or changing scientific ideas into really useful products for industry, medicine or the developing world, for example. You will discover some of the many things we can do with science when we take it 'out of the classroom' and use it to solve a wide range of practical problems.

A 'Questions' section at the end of each article lets you check how well you understood the article. Another section called 'Activities' suggests topics that you can discuss with other students, or ideas you can use to help you find out or understand more about the article you have just read. Sometimes there will be opportunities to talk about how science can really change people's lives, for better or worse, and to discuss whether new developments are a good idea or not.

As you look through *The Amazement Park* you will see the page numbers are in different colours. This is because the articles are written in three slightly different reading styles; the style of the red pages is slightly harder to read, and the style of the green pages is slightly easier, so you can choose the topic and the reading style that you like best.

Information for teachers

The articles in *The Amazement Park* all link to topics covered in the Year 8 Programme of Study from the QCA Scheme of Work. They are written at a conceptual level appropriate for Year 8 students and, within each Unit, the articles provide progression by reading age. *The Amazement Park* is ideal for the 'Ideas and Evidence' part of your programme of study, supporting development of all the relevant Scientific Enquiry skills needed.

On the most simple level, a 'Questions' section at the end of each article allows you to check students' understanding of the article. On a more complex level, the 'Activities' section allows students to explore, through discussion, research and role-play, their responses to issues raised, and their opinions about how, when and why science discoveries should be used, often linking with Citizenship issues.

There are three articles linking to each QCA Unit in the Scheme of Work, covering different topics, allowing pupils to read different articles and then report back to the class, if you wish.

You're never going to eat that!

Setting the scene

You know that we need to eat a balanced diet to stay healthy. Packets of food are labelled with 'nutritional information' to help us make the right choices, and 'sell by' dates to make sure we eat them at the right time. But what do these dates really tell us, and does it matter if we ignore them?

Chuck it out

Foods that keep for a long time, such as pasta or biscuits, usually have a 'best before' date. Although these foods don't usually 'go off', they do lose their flavour after a while, so they taste stale. Foods that will 'go off' have a 'sell by' date on them. If you eat them after this date, you may get food poisoning. Foods that go bad very quickly have a 'display until' date, to tell the shopkeeper when to sell them by, and a 'use by' date to tell you when to eat it by. These foods also have instructions telling you to keep them in the fridge or freezer.

You wouldn't want to eat this would you, but how can you tell if food is just a little bit 'off'?

There's no accounting for taste!

The Chinese have a favourite delicacy called 'thousand year old eggs'. Duck eggs are buried in soil for about 3 or 4 months then dug up and eaten! The yolk goes green and the white goes black with yellow or orange flecks. Fans of this type of egg say it has a strong flavour a bit like cheese. Many people like to eat blue cheese, which is cheese with blue mould growing in it.

Poison!

Food poisoning can make people very ill; sick people, the very elderly or very young children can even die. Usually it happens because someone eats some food that has 'gone off' – it has started to rot. Bacteria from the bad food breed inside the person, causing vomiting and diarrhoea. Most food poisoning is caused by undercooked meat or eggs, or by cooked food that has been stored too long. Cooking food thoroughly just before eating it, rather than just warming it up, kills bacteria in the food. Food that has 'gone off' often tastes bad, because lots of the bacteria give off poisons that taste horrible. It is our body's way of warning us not to eat the food.

It's gone mouldy!

Years ago, a loaf of bread would only last two or three days before it started going mouldy. Now loaves may last more than a week. Modern bread has chemicals called preservatives in it to stop bacteria growing. Some people say manufacturers should not put so many preservatives in food, because the chemicals are bad for us.

Won't it glow green?

Some foods are sterilised using ultra-violet radiation or radioactive radiation, which both kill bacteria. Some people don't like this because they worry that the radiation will stay in the food, but it doesn't. Some scientists say it is safer than using preservatives because it doesn't leave chemicals in the food.

That sucks!

Vacuum packaging is when the food is packed in plastic wrapping and then the air is sucked out. This makes the food last longer because lots of the bacteria that make food go off are carried around in the air. Keeping out the air keeps out the bacteria.

Vacuum packaging makes food keep longer, but it doesn't keep for ever.

What a stink

Bad food smells horrible because the bacteria give off smelly gases as they make the food go bad. Recently some scientists have discovered a use for these gases. They put a label, covered with special dyes, in with the food. When the gases touch the dye they make it change colour, so the label changes colour. People can look at the colour of the label to tell if the food is still fresh.

- What type of foods have a 'best before' date, and what is it for?

- How can you kill the bacteria that cause food poisoning?

- How does our body warn us not to eat food that is going bad? Do you think this warning system always works?

- People in the catering trade are always taught that you must never store raw meat where it can drip onto cooked meats. Can you explain why?

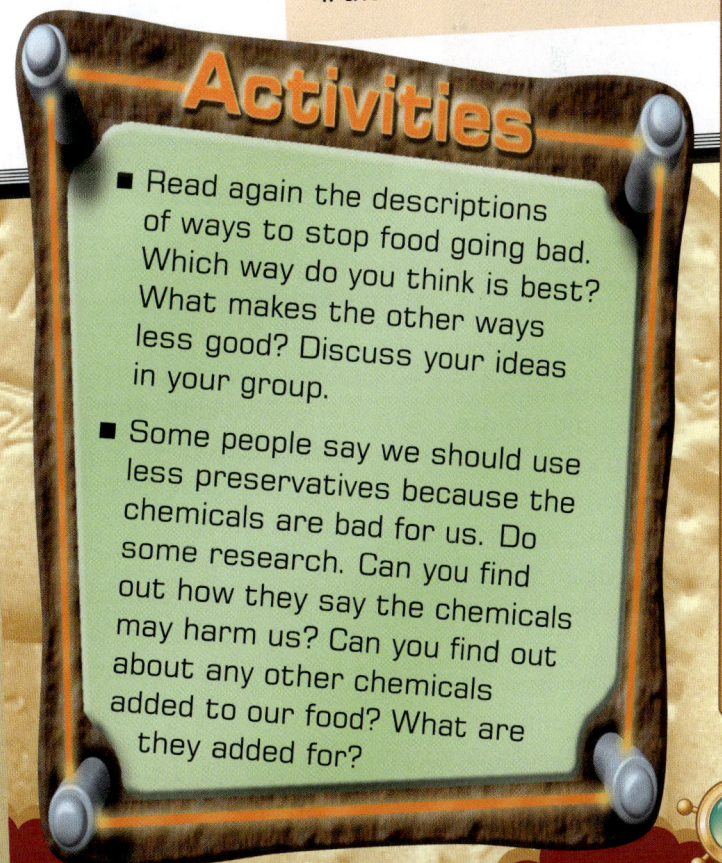

Activities

- Read again the descriptions of ways to stop food going bad. Which way do you think is best? What makes the other ways less good? Discuss your ideas in your group.

- Some people say we should use less preservatives because the chemicals are bad for us. Do some research. Can you find out how they say the chemicals may harm us? Can you find out about any other chemicals added to our food? What are they added for?

Eat your greens

Do you ever get fed up with parents saying, 'Eat this, it's good for you' or 'Don't eat that, it's bad for you'? Parents forget that as you grow up, you learn to make sensible choices for yourself! Read these newspaper reports showing how hard it can be for parents to make the right choices for their children, then think 'What would I do?'

Setting the scene

Is your toddler malnourished?

 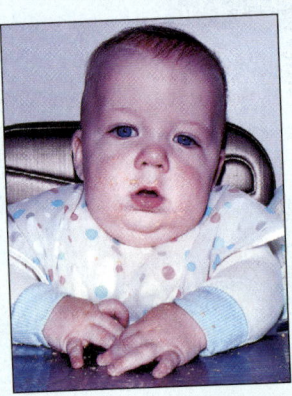

Doctors tell us that both these children could be malnourished! Like us, you probably thought that children were only malnourished if they did not get enough to eat. Now doctors tell us that young children can be overweight and malnourished at the same time! 'Hundreds of children in rich countries are getting too many calories and not enough vitamins and minerals,' one paediatrician told us. 'We're not saying don't give your toddler chicken nuggets and chips,' she told us, 'Just make sure it is part of a balanced diet, with lots of fresh fruit, vegetables and fruit juice, so they get the vitamins and minerals they need for healthy growth.' Then she added, 'And make sure your toddler gets lots of active play. Play helps them develop coordination, and stops them getting obese.'

Muesli belt malnutrition

Doctors in America are worried by the rise in what they are calling 'muesli belt malnutrition'. Obesity in America is a BIG problem, we are told, and leads to lots of illnesses, such as heart disease, high blood pressure and even cancer. Some well meaning parents, hoping to protect their children from these problems in later life, are putting their babies and toddlers on the sort of low-fat, high-fibre diets that are excellent for adults. 'Young children have different nutritional needs from adults,' our paediatrician explained, 'Up to half a baby's or toddler's calories should be from fatty foods. They are growing fast and they are usually much more active than adults. An average toddler may walk or run the equivalent of more than 15 kilometres in a day. Also fats are necessary for the brain to develop properly.' 'So how do we know how much fat to give toddlers?' our reporter asked. 'A toddler should have three cups of milk a day, or the equivalent in yoghurt or cheese, and they should have two servings of meat, peanut butter, eggs or beans, as well as plenty of fruit, vegetables, bread and cereal.'

Doctors say British toddlers are fine!

BRITISH RESEARCHERS SAY THAT THERE IS NO EVIDENCE THAT BRITISH TODDLERS are suffering from 'muesli belt malnutrition'. 'Most toddlers in Britain eat an adequate diet,' our reporter was told, although the paediatrician went on to tell him that a study of toddlers from Bristol showed that those who ate least fat are more likely to have insufficient vitamin A, found in milk and cheese. Those that ate most fat were more likely to be short of iron and vitamin C, because their intake of fruit and fruit juice was low. 'Fortified foods, such as breakfast cereals, have had extra vitamins and minerals added and they are an excellent way to help young children get enough vitamins and minerals,' she added.

Qs

- Re-read the section 'Is your toddler malnourished?' Then explain in your own words what you think 'malnutrition' is.

- How much of a baby's or toddler's calories should come from fatty foods?

- Why do you think there is a connection between low-fat diets and low levels of vitamin A?

- What do you think the sentence 'An average toddler may walk or run the equivalent of more than 15 kilometres in a day.' means?

Activities

- Discuss how a family with a toddler could plan their meals and snacks so that both the adults and the toddler eat healthily.

- Some surveys have suggested that too many vitamins and minerals can be harmful, as well as too few. Discuss how you think this affects the way in which foods are 'fortified'.

- Design a poster for a health clinic, to advertise a healthy diet for toddlers.

Is global warming made worse by farting?

You know that a balanced diet provides all the raw materials our body needs, but you also know that we need fibre in our food, to keep our digestive system healthy. So what happens to the fibre that we can't digest, does the amount of fibre we eat affect us, and what about animals that have a diet that is almost all fibre? Read these newspaper reports to find out more.

WHO DO YOU SHARE YOUR FOOD WITH?

YOU PROBABLY THOUGHT you were the only one to eat your food. Not so, dieticians tell us! Apparently our intestines are full of bacteria that digest the fibre that we can't digest. These bacteria are actually helpful because they change some of the fibre into substances that we can absorb, but they also give off methane gas as a by-product of their digestion. This gas passes out of the body as 'wind'. Some people pass wind up to 40 times a day, but the average seems to be about 15 times a day.

Is global warming made worse by farting animals?

COWS AND SHEEP could make global warming worse, scientists from Australia are telling us. Their very high-fibre diet means that sheep give out about 25 litres of methane per day, and cows give out a staggering 280 litres of methane per cow per day! Methane is a greenhouse gas that contributes to global warming. The scientists tell us that cows and sheep account for about 14% of the greenhouse gases from Australia and 50% of the greenhouse gases from New Zealand.

Farming kangaroos instead of sheep and cows could solve the problem, because kangaroos have bacteria in their intestines that change the gases into acetate, which the kangaroos use as an energy source. Scientists are now wondering if genetic engineering could be used to successfully introduce the same bacteria into cows and sheep. It would also mean cows and sheep could survive on much less grass than they need at the moment.

25 LITRES OF METHANE PER SHEEP PER DAY. 50 MILLION SHEEP. THAT'S E A LOT OF METHANE.

Is panda poo the fuel of the future?

THE GIANT PANDA FROM CHINA EATS BAMBOO and virtually nothing else, and bamboo isn't very nourishing – it's like eating garden canes from the garden centre!

A researcher from Kitasato University in Tokyo in Japan wondered how the pandas survive. He decided that in their digestive system they must have really powerful bacteria that digest the bamboo and change it into something that the pandas can get energy from. So he decided to have a look. When he examined panda poo, he found five different types of bacteria from the panda's digestive system that seemed to be particularly powerful. In 17 weeks the bacteria changed 100 kilograms of vegetable waste to gases, leaving only 3 kilograms of residue. Each kilogram of waste gave 100 litres of hydrogen that could be used in fuel cells to drive cars or other vehicles.

There is more work to be done, but possibly in the future we will all drive cars powered by kitchen waste and the bacteria from panda poo!

Qs

- What useful purpose do the bacteria in our intestines serve?

- What type of diet causes the bacteria to produce most gases?

- Suggest *two* possible ways in which Australia and New Zealand could reduce their emissions of greenhouse gases.

- Use a diagram or a word equation to show what happened to the vegetable waste that the Japanese researcher used.

Activities

- The Japanese researcher showed that it was possible to make hydrogen fuel using the bacteria from a panda's digestive system. Do you think pandas are the only animals that could be used? Discuss which other animals you might investigate? Why would you choose these animals?

- Cows give off methane gas as a by-product of their digestion. Find out some of the things methane gas can be used for. Then design a way to collect and use the methane gas from a herd of cows.

A little of a bad thing does you good!

You have learned that our lungs take oxygen from the air, and our blood carries it round our bodies to where it is needed. We need oxygen to stay alive – that's why firemen wear breathing apparatus when they go into smoky buildings. Our interviewer talks to Dr Addison, who apparently has a surprise for us about one of the most dangerous gases in smoke.

Setting the scene

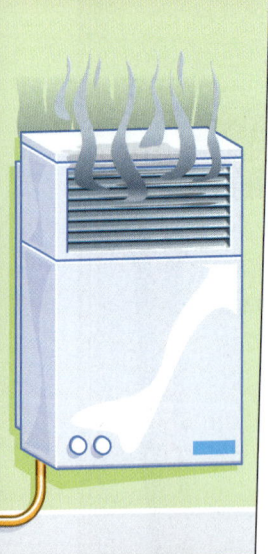

Interviewer Dr Addison, I'm told you are an expert on carbon monoxide. So what is it?

Dr Addison It's a colourless, odourless gas. That means you can't see it and you can't smell it.

Interviewer But if we breathe it in, it is very dangerous?

Dr Addison Oh, yes, very dangerous. Our blood is better at carrying carbon monoxide than it is at carrying oxygen. So, if you breathe in carbon monoxide it will get carried round your body instead of the oxygen. Your cells won't get enough oxygen and you will die.

Interviewer But if we can't see it or smell it, how can we avoid it?

Dr Addison It is quite rare. It comes from car exhausts, and heaters or central heating boilers that aren't working properly.

Interviewer That sounds scary.

Dr Addison It's not as bad as it sounds. If a boiler or heater isn't burning properly there will be black or yellowish sooty marks on the wall above it. Never use a heater that looks like it is smoky.

Interviewer How could we tell if someone had carbon monoxide poisoning?

Dr Addison A heater or fire would be on somewhere nearby. They would seem sleepy and confused. They might complain of a headache, or feeling sick or dizzy. Turn the heater off immediately, open the doors and windows and call a doctor or ambulance.

Interviewer What's the surprise you have for us?

Dr Addison Carbon monoxide is not always harmful. It can be used to help heart attack patients.

Interviewer That doesn't sound very likely. Tell us more.

Dr Addison All the cells in our bodies actually make very tiny amounts of carbon monoxide. It is used to control parts of our blood system. For example, small amounts of carbon monoxide help blood vessels to open, making blood pressure lower, and helping people who suffer from high blood pressure.

Interviewer So how can carbon monoxide help heart attack patients?

Dr Addison When someone has a heart attack, the heart muscle doesn't get enough oxygen and it can be permanently damaged. Research has shown that small amounts of carbon monoxide stop the damage to the heart muscle.

Interviewer So breathing carbon monoxide will cure a heart attack?

Dr Addison No, certainly not! It has to be really tiny amounts of carbon monoxide, and it has to be given in a special medicine form that releases the carbon monoxide exactly where it is needed.

Interviewer So does carbon monoxide have any other amazing uses?

Dr Addison We are not sure yet, but doctors have made carbon monoxide medicines that can be dissolved in water. This makes it easier to give the right dose of carbon monoxide in the right place. Doctors hope that one day they will be able to use carbon monoxide to help people who have had strokes, and people with brain injuries.

Interviewer Well, thank you for talking to us. I have learned a lot.

Qs

- Explain what a colourless, odourless gas is.

- Why is it dangerous to breathe in carbon monoxide gas?

- List some things that a person might feel, or complain of, if they had carbon monoxide poisoning.

- How can carbon monoxide help a person who has had a heart attack?

Activities

- Suppose you went to stay in holiday accommodation. It was cold, but the only heater had sooty stains above it. Discuss all the things that you would do. Think about the safety of other people, as well as your own safety.

- Every room with a gas fire must be fitted with an air vent, to make sure the gas fire has enough oxygen to burn properly. Design a short radio 'safety information bulletin' explaining why it is dangerous to cover this air vent.

Breathing without air

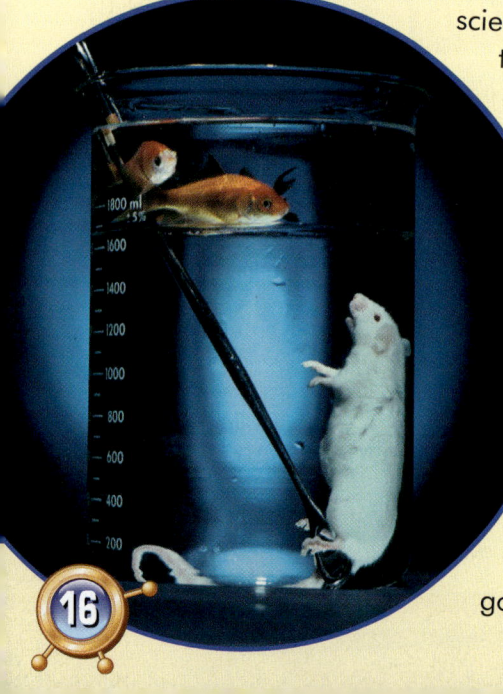

You know that we need air to breathe. Our lungs take oxygen from the air, and our blood carries it all round our body, to all the cells. Waste carbon dioxide is carried back to the lungs in the blood, and we breathe it out. But some doctors have said that we don't need to breathe air at all! Read on to find out more.

Humans, and all other mammals, need to breathe air to stay alive, but have you ever wondered why fish can stay under water and you can't? The answer is easy – fish have a way of breathing under water and you don't. What that really means is that fish can get the oxygen they need from water, but we have to get our oxygen from air. So some scientists started wondering, 'Is there any way humans could get the oxygen they need from water, or from some other liquid?' And, guess what? They tried to find out!

The first problem – not enough oxygen

The first problem with humans breathing under water is that there is not enough oxygen in water. That doesn't matter for fish, because they are cold blooded. Their bodies do not work as fast as ours do, so they do not need so much oxygen. So scientists had to find a liquid that could hold much more oxygen than water could. In the 1960s they found a way to dissolve lots of oxygen in salt solution. The next task was to see if lungs could push the salt solution in and out. But they didn't want to test it on humans, in case it didn't work.

The second problem – the carbon dioxide won't come out!

Rats and mice could breathe the salt solution and oxygen quite successfully, and it gave them the oxygen they needed. But the salt solution was nowhere near good enough at holding carbon dioxide, so a lot of the waste carbon dioxide stayed in their bodies. And too much carbon dioxide is poisonous! The scientists had to find a different liquid, one that was good at holding carbon dioxide as well as oxygen.

The third problem – they damage the lungs

In 1966, scientists thought they had solved the problem. They found that liquids called fluorocarbons were very good at dissolving oxygen and carbon dioxide. Experiments showed that rats and mice could breathe the liquid successfully, but it damaged their lungs. Scientists weren't sure why; perhaps there were toxic chemicals in the fluorocarbon liquids. Whatever it was, the fluorocarbons clearly couldn't be used for humans.

We can test it on humans at last!

In the 1990s doctors finally found a liquid that they decided was safe to test on humans. It was a chemical called perfluorooctyl bromine, usually called perflubron, and it was already being used as a blood

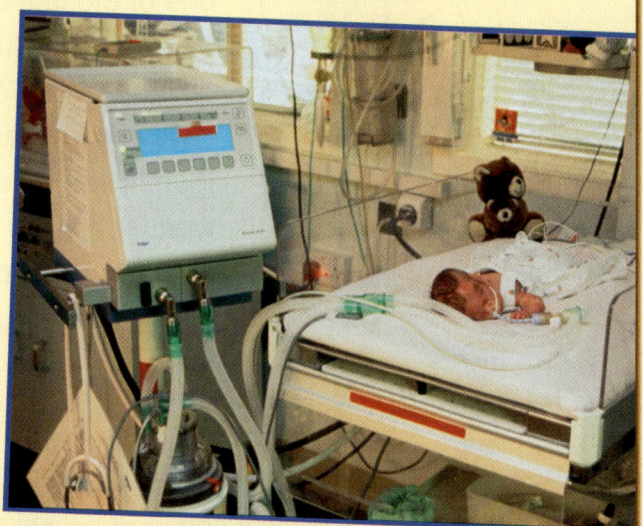

substitute. They tried out the new liquid on patients whose lungs had collapsed because of infection, burns or breathing smoke, and on very premature babies whose lungs were not working. At the time, ventilators were being used to pump air into the lungs of these patients, but often the pressure needed to get air into their lungs caused permanent lung damage, which made the patients die anyway. The perflubron liquid filled their lungs and allowed oxygen into and out of the lungs without damaging them. The tests worked so well that the liquid was given a new name, LiquiVent, and it is still being used in hospitals today.

Babies whose lungs are too delicate to breathe air, can survive if they breathe LiquiVent.

Qs

- Why can fish get enough oxygen from water but we can't?

- Why was salt solution with oxygen in it an unsuitable liquid for breathing?

- Explain why ventilators often damaged lungs.

- List the three different names for the chemical that doctors found could be breathed successfully by humans.

Activities

- Is it acceptable to test new treatments on animals first? Discuss what you think in your group.

- Imagine you were offered a new treatment that hadn't been tried on humans before. Would you accept it? What would you think about before you made up your mind?

Can worms prevent asthma?

Setting the scene

Most of us take breathing for granted, but if you are one of the one in eight young people in the UK being treated for asthma symptoms, you will certainly not take breathing easily for granted. Recent surveys have shown that more children in the UK have asthma than in any other country in Europe, and you may well wonder why no cure for asthma has yet been found. Dr Frank explains some of the problems.

Interviewer Hello, Dr Frank. We are hoping you are going to answer some very difficult questions for us, this morning.

Dr Frank I'll do my best.

Interviewer A generation ago most people hadn't even heard of asthma, because it was so rare. Now one in eight children in this country has asthma. So how does asthma affect people?

Dr Frank It depends on how well they are treated. Many people cope extremely well with their asthma, and lead perfectly normal lives. Many top athletes have asthma – Paula Radcliffe, the world record-breaking marathon runner, and Paul Scholes, the England and Manchester United footballer, for example. As Paula says 'If you learn to manage your asthma and take the correct medication, there is no reason why you shouldn't be the best.'

Interviewer So what's the cure for asthma?

Dr Frank We don't know yet. We can't find a cure until we know what causes asthma. Smoking, owning pets or being allergic to dust mites all seem to make it more likely that someone will have asthma, but we don't know if they actually cause asthma. For instance, it might be that cigarette smoke kills a type of bacteria that makes our immune system strong and stops us getting asthma. We have to keep doing research to test these ideas.

Interviewer I think I understand. Are you saying that when there seems to be a link between asthma and something else, you have to check to see if it really is a link or if it is coincidence?

Dr Frank Yes, that's a very good way of putting it.

Interviewer Tell us about some of the latest studies about asthma, please Dr Frank.

Dr Frank Well, scientists in Belgium have found a closer link between asthma and how much time children spend at indoor swimming pools than between asthma and smoking or asthma and owning pets.

Interviewer So there's a link between asthma and swimming?

Dr Frank We're not sure yet. But sweat and urine both react with the chlorine in swimming pools to make a gas called nitrogen trichloride that may cause asthma. The scientists are calling for swimming pool water to be sterilised using ultra-violet light instead of chlorine.

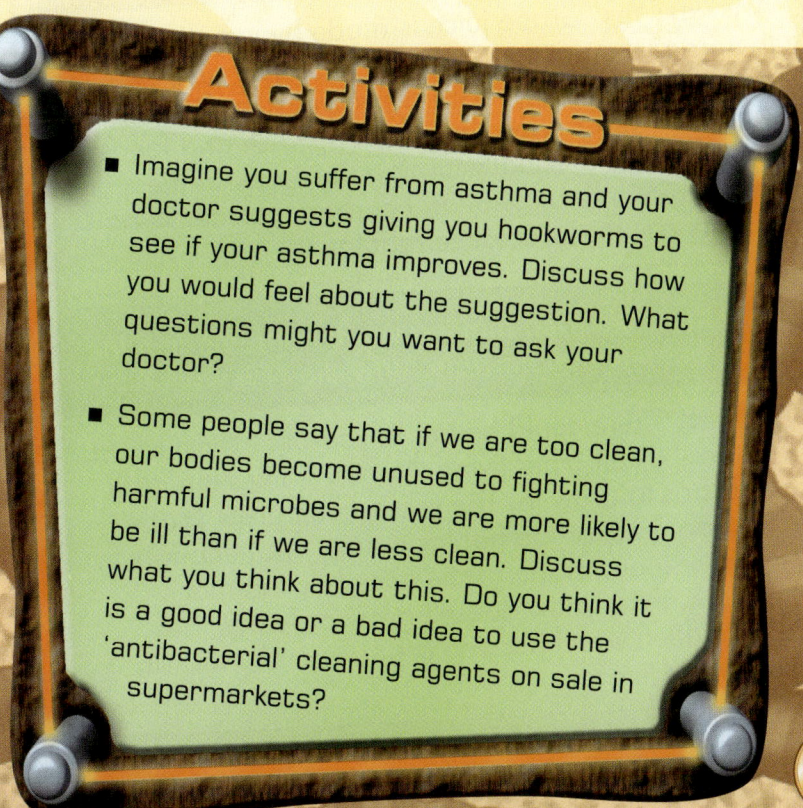

This looks horrible, but it is possible that it stops people getting asthma.

Interviewer What other studies are there?

Dr Frank A study of people in towns and remote villages in Ethiopia has shown that the people in the villages have hookworms, but they are less likely to have asthma.

Interviewer Couldn't it just be that people in towns are more likely to breathe traffic fumes?

Dr Frank That's what made the Ethiopia study so interesting. There just didn't seem to be enough traffic in the towns to account for the difference in the number of people with asthma. Surveys in the UK have also shown that the increase in asthma happened at the same time as the decrease in hookworms. Scientists are wondering if the hookworms somehow make people less likely to have allergies or asthma. They suggest that perhaps doctors should give hookworms to children with asthma to see if their asthma improves.

Interviewer Well, I wonder what our listeners with asthma would think of that? Thank you for talking to us, Dr Frank.

Qs

- What proportion of children in the UK suffer from asthma?

- List three things that are linked to an increased risk of having asthma.

- Explain how scientists think indoor swimming pools may cause asthma.

- What two sets of data make scientists think there may be a link between asthma and hookworms?

Activities

- Imagine you suffer from asthma and your doctor suggests giving you hookworms to see if your asthma improves. Discuss how you would feel about the suggestion. What questions might you want to ask your doctor?

- Some people say that if we are too clean, our bodies become unused to fighting harmful microbes and we are more likely to be ill than if we are less clean. Discuss what you think about this. Do you think it is a good idea or a bad idea to use the 'antibacterial' cleaning agents on sale in supermarkets?

Moody microbes

Micro-organisms are tiny living things, far too small to see without a microscope. You know that some micro-organisms are harmful – they cause diseases or make food go bad – but others are useful – we use them for making bread, yoghurt, wine and other foods and drinks. Find out how this expert from a wine company uses micro-organisms, and some of the problems he has.

Question: I have seen grape juice on sale in supermarkets. Is there any difference between grape juice and wine?

Answer: Yes. Grape juice is made by just squashing the grapes to get the juice out. Wine is made by adding a micro-organism called yeast to the grape juice. The yeast grows in the grape juice and feeds on the sugar, gradually turning the sugar in the grape juice into alcohol and carbon dioxide. Wine makers call this process fermentation.

Question: We tried making some homemade wine once, but it tasted horrible. Why didn't it work?

Answer: There could be lots of reasons. Yeast is just the same as all other living things – it likes the right conditions to grow in. The yeast won't grow properly if it is too cold, or too hot, or it hasn't got enough sugar for food, or the grape juice has too much acid in it. If the yeast doesn't grow properly, the fermentation won't happen properly and your wine won't taste like wine.

Have you ever tried making wine? Do you know why it is so hard to get it tasting really good?

Question: *Your company has just won a Good Wine Award. Why do people like your wine so much?*

Answer: We have market researchers who go out onto the High Street and find out what people want their wine to taste like. Then we change the way we make our wine, so that it tastes like people like it. We can change things like the type of grape we use, the temperature and the acidity of the mixture and how long we let the wine ferment for.

Question: *How can you tell before it is finished, whether a wine is going to be good or bad?*

Answer: Years ago we used to have wine experts who tasted the wine at different stages when it was being made. They knew what it should taste like at every stage. Now we use scientific tests to measure things like how much the yeast has grown, and how much sugar or acid there is.

Question: *One final question – has the way your company makes wine changed at all?*

Answer: Yes, now we do lots of scientific tests on our wine. Computers monitor our wine all the time while it is being made, to check that the yeast has the best conditions for growing in. Because we check our wine all the time, we are able to make small changes straight away, to make sure our wine always comes out tasting the same. So when people buy our wine they know it will taste just as good every time. It doesn't have 'good years' and 'bad years' like old-fashioned wines used to.

Why do all these wines taste different? Can you suggest some reasons why?

- What are micro-organisms?

- What does the yeast in wine feed on? What does it turn it into?

- List three things that might stop yeast growing properly.

- List three things that wine makers might measure while their wine is being made.

- Can you suggest a possible reason why using scientific tests and computer monitoring makes a better wine than tasting it at different stages?

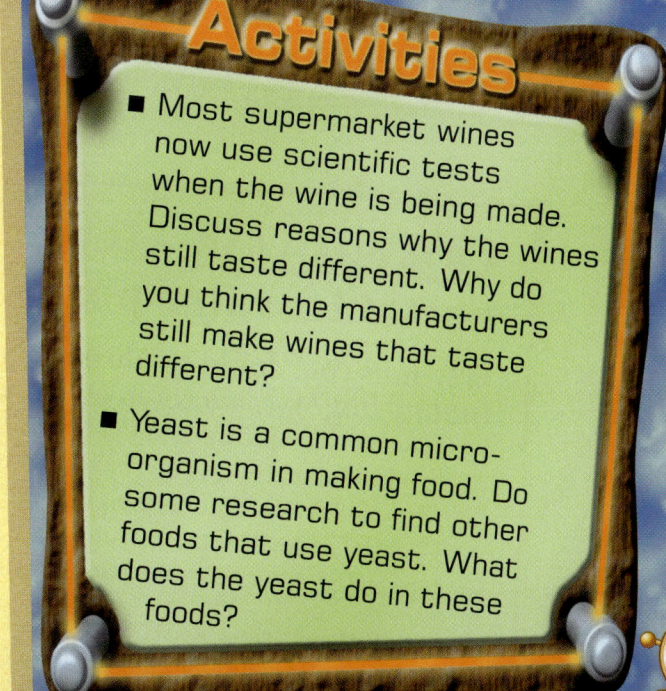

Activities

- Most supermarket wines now use scientific tests when the wine is being made. Discuss reasons why the wines still taste different. Why do you think the manufacturers still make wines that taste different?

- Yeast is a common micro-organism in making food. Do some research to find other foods that use yeast. What does the yeast do in these foods?

'Water, water everywhere, but not a drop to drink'

You know that there are many different types of micro-organisms, and some of them harm us by causing diseases. One of the most common ways harmful micro-organisms get into our bodies is through drinking dirty water. We are lucky to have clean water to drink, but how did people manage before water was piped in water mains and what alternatives were there?

Setting the scene

Until modern water purification was invented, people could catch diseases and die if they drank water. Most Western cultures drank beer instead, which was safe because the water was heated during the beer making process. Many Eastern cultures drank tea, which was safe because the water was boiled to brew the tea. Even today, these historical differences are still visible, people from Eastern cultures are often less able to tolerate alcohol than people from Western cultures.

Read these advertisements to find out about some of the choices available for making water safe to drink today.

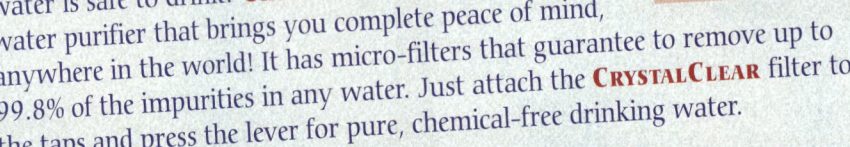

Sparkle – the purifier you can trust!

Forget inconvenient filters or chemicals – **Sparkle** uses the power of natural radiation to clean the water coming to your tap. Our purifier harnesses the natural power of ultra-violet radiation to kill bacteria and viruses in the water. The **Sparkle** unit is plumbed in to your water supply, so that all your water flows past powerful UV lamps, giving you up to 50 litres a minute of chemical-free, no hassle, pure water.

The latest **Sparkle** model has an automatic water cut-off switch for when the UV lamp stops working, as well as the latest easy-change bulb.

Sun Power – the solar distillation unit

Perfect for use in developing countries. Sun Power needs no electricity, no pumps and it works for ever! Just fill the tray with dirty water

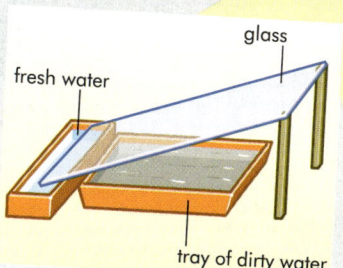

and let the Sun do the rest. The heat of the Sun evaporates the water from the tray, and it condenses on the glass plate, then runs down the plate and collects in the fresh water drain. Sun Power can be used anywhere sunny to provide clean, safe drinking water for families or communities. Each square metre of glass plate provides about 1 litre of drinking water per hour in the sunshine.

Saris stamp out cholera!

Communities in Bangladesh are using old saris to beat cholera. Women in the villages have halved the number of cases of cholera just by filtering drinking water through folded saris. The cholera is caused by bacteria in the water, but the bacteria cling to tiny plants in the water called plankton. The folded saris filter out the plankton and trap 99% of the cholera bacteria. This is what one Bangladeshi woman told us, 'Our children don't get sick nearly so often. We want to tell every village in Bangladesh and India how they can have clean water too.'

Qs

- Only one of the water purification methods needs an electricity supply. Which one?

- CrystalClear claim you can use their water purifier 'anywhere in the world'. Do you think they are right?

- The Sun Power unit does not provide endless water. What things do you think it might still be safe to do with unpurified water?

- Communities in Bangladesh found that folded saris are much better at filtering water than unfolded saris. Suggest a reason why.

Activities

- The CrystalClear portable water purifier claims to remove up to 99.8% of impurities from water. Discuss whether or not it is sensible to use the water from the CrystalClear water purifier to mix baby food.

- Sparkle claims to use natural ultra-violet radiation to kill micro-organisms. Do you think this claim is reasonable? Do some research to find out more about ultra-violet radiation. Tell the rest of your class what you have found out.

Are hospitals safe?

Setting the scene

Although our bodies have natural barriers to stop micro-organisms getting in and our immune systems fight micro-organisms that do enter, sometimes harmful micro-organisms still enter our bodies and make us ill. Our experts explain how we can fight harmful micro-organisms, and about the latest efforts of scientists and doctors to deal with powerful 'superbugs'.

Three lines of defence

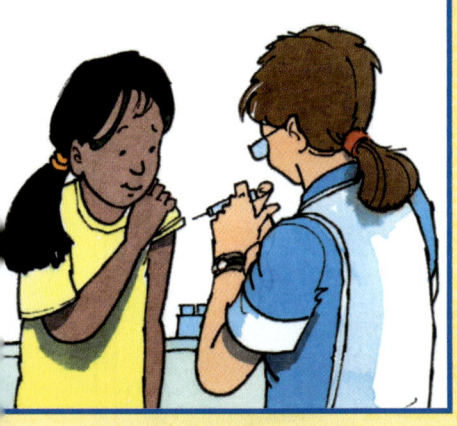

Vaccines protect us from many harmful diseases, such as cholera, typhoid, measles, mumps and tetanus.

Our first line of defence against micro-organisms are the antiseptics, disinfectants and antibacterial cleaners. These chemicals kill micro-organisms before they enter our bodies.

Vaccines are the second line of defence. These medicines, often given by injection, fool our body's immune system into making its own antibodies against harmful micro-organisms. Then when the real micro-organisms enter our body, our immune system is ready for them, with the right weapons to kill them.

The third line of defence are antibiotics. These are medicines given to someone who is already ill, to kill the bacteria in their body. Antibodies only kill bacteria, so they won't cure diseases caused by viruses, such as common colds or flu.

What are 'superbugs'?

Some bacteria are very hard to kill with antibiotics. These bacteria are often known as 'superbugs'. The first antibiotics were discovered in the 1930s and they were very effective, causing 'miraculous' recoveries from many diseases. Unfortunately bacteria evolve and change very quickly, so after a few years bacteria evolved that were immune to some antibiotics. Now there are some 'superbugs' that are resistant to nearly all our antibiotics. Doctors prescribing antibiotics always explain how important it is to finish all the course of antibiotics even if you feel better. This is because you start to feel better before all the bacteria have been killed, but if you stop taking the antibiotics the remaining bacteria will grow again, and may have become resistant to the antibiotic you were prescribed.

The race to beat the 'superbug'

The race is really on to beat the 'superbugs' before they beat us. 'Superbugs' cause 5000 deaths per year in British hospitals alone, and many more people are disfigured by infected wounds that will not heal, causing serious scarring or even amputation. Many of these cases could be avoided. Doctors are being urged to avoid prescribing antibiotics wherever possible, so bacteria are less likely to become resistant to them, and to improve their hygiene, washing their hands thoroughly with antibacterial chemicals before touching patients, for example.

The new generation of antibiotics

Can this crocodile provide the antibiotics of tomorrow?

There are some plant and animal compounds that look like they might provide the antibiotics of tomorrow. Garlic and tea tree oil have both been used in medicine for hundreds of years. During the Second World War, Australian soldiers called tea tree oil a 'first aid kit in a bottle' because it was so good at fighting infections. Scientists have now found substances in both garlic and tea tree oil that kill most of the known 'superbugs'. More tests are still being carried out.

In the late 1990s, a BBC science correspondent filming crocodiles in Australia noticed that although the crocodiles often fought and injured each other, the wounds never became infected, even though people's cuts and scratches nearly always became infected if they entered the water where the crocodiles lived. Scientists began investigating why. In 2000, scientists in America identified a powerful antibiotic in crocodile blood, which kills 'superbug' bacteria resistant to all the standard antibiotics. They have called the substance 'crocodillin'. They hope that soon tests will be completed so that crocodillin can be used to treat human infections.

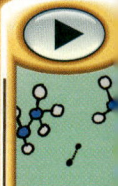

 **Qs**

- Why did antibiotics gradually become less effective after they were discovered?

- Explain in your own words why scientists are saying doctors should prescribe less antibiotics.

- You are more likely to 'catch' a resistant 'superbug' in hospital than at home. Suggest some possible reasons why.

- 'Even if we get new antibiotics that work against superbugs, it won't be safe to relax.' What do you think this doctor means? Do you agree with her?

 **Activities**

- Some people worry that using too many antibacterial cleaning products will encourage resistant 'superbugs' to spread. Discuss what you think.

- Tea trees are under threat because the oil from them is valuable. Discuss what, if anything, we should do about this.

- Many people feel 'let down' or disappointed if a doctor won't prescribe antibiotics to help them get over an illness quickly. Design a simple public information poster for a doctor's waiting room, to help explain why sometimes the doctor won't prescribe antibiotics.

Jumbo
landscape designers

Humans often do things to change the environment. Sometimes they make it better for humans or for animals and sometimes they cause damage. But are humans the only animals that change their environment? And how can we tell whether a change is good or bad? Read the diary of this wildlife ranger from Kenya, to find out more.

Tsavo National Park, 1969

I have just come to work as a wildlife ranger in Tsavo National Park. I feel very lucky, it really must be the most beautiful place in the whole of Africa. I am very worried about the Park though, the elephants seem to be wrecking it. They are pushing over small trees and trampling large holes where the roots used to be. There are more elephants in the Park than there used to be because there are more villages round the edge of the Park than there used to be. The elephants try to stay away from the villages, so they are moving into the Park. I think we should shoot some of the elephants to save the Park, but the Kenyan Government won't let me.

Tsavo National Park, 1970

It has all happened so quickly. A few months ago there were too many elephants, now so many of them are dead. We are having a terrible drought here, and the elephants are starving. The old female elephants, that lead the herds, get weak and won't go far from water. Whole families just stay by the water and half of them starve. I don't think any baby elephants have been born this year.

Tsavo National Park, 1973

The Park just looks so good, better than ever! New trees are springing up all over it — their seeds were carried round in the elephant dung before the drought. Most of the holes that the elephants trampled are now water holes. I've seen more species of birds and animals round them this year than ever before. There's more open space round the water holes too, which means more grazing animals are able to find food.

Tsavo National Park, 2003

I can't believe I have now been ranger here for more than thirty years. I have seen so many changes. The huge numbers of elephants when I first came here, and the drought of 1970, which seemed such disasters at the time, but now I can see what a good thing they were. They were all just part of a natural cycle. The elephants actually made the habitat that so many different animals need. When I look round the Park now I can see that the areas with the largest number of different species are actually the areas the elephants seemed to be wrecking all those years ago!

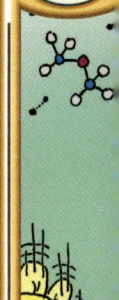

Biodiversity

The number of different species living in a habitat is called the 'biodiversity'. Scientists use the biodiversity of a habitat to tell how healthy an environment is. If the environment is healthy, it has a lot of different plant and animal species living in it. A damaged environment has fewer species in it. The elephants in the Tsavo National Park made the environment there healthier because what they did to it meant more species of animals could live there.

Qs

- Why did the ranger think the elephants were wrecking the Park in 1969?

- Describe two ways the drought in 1970 affected the elephants.

- Give two changes the ranger noticed in the Park in 1973.

- By 2003, how had the ranger changed his mind about the damage the elephants had done?

Activities

- Discuss how we might be able to tell if a change is good for an environment or not.

- Discuss what things scientists need to know about an environment before they can protect it.

- Imagine you are a wildlife ranger teaching children in an African primary school to look after their environment. Write a short play, that two or three people could perform, to show the children how wild animals could be useful to them.

Use a pest to catch a pest!

The animals in an environment eat the plants, and each other! Sometimes 'pest' species, such as greenfly or locusts, seem to suddenly appear and eat huge amounts of crops. So why does this happen, and what can farmers or gardeners do about it? Read these questions and answers from the pages of a farming magazine.

Setting the scene

Question: *I grow roses on my smallholding, to sell to florist shops. This year my rose bushes are covered in greenfly. Why are they attacking my roses, and what can I do about it?*

Answer: Your roses are the perfect place for greenfly to live! Your roses provide them with lots of food to eat and hardly any predators to eat them! So they breed very quickly and the numbers increase until your roses are covered in them. There are two ways to get rid of your greenfly; you can spray the roses with an insecticide chemical that kills them or you can find some predators to put on them. You can buy ladybird grubs that grow into ladybirds and eat the greenfly, but you may have left that a bit late for this year.

Question: *I have heard that you can use predators to control pests. How do you choose the right predator?*

Answer: It can be really difficult to choose the right predator, and in the past scientists have often got it wrong. They have to make sure the predator only eats what they want it to eat, not other things as well. In the French Polynesian islands, farmers were having trouble with imported African land snails eating their crops, so scientists imported another snail, the *Euglandina* snail, to eat the land snails. It didn't eat the land snails, but it did eat the native snails and made 56 out of the original 61 species of native *Partula* snails extinct. Now scientists try not to introduce predators from other countries, because they can never quite be sure what they will eat.

Question: We are turning our farm into an organic farm, so we won't be able to use pesticides. I have heard that we can use beetle banks. What are they, and how do they work?

Answer: Beetle banks are strips of long grass, one or two metres wide, that farmers leave growing across the middle of their fields between the crops. Lots of different species of beetle eat the pests that feed on crops; the long grass gives the beetles somewhere to breed, so they spread over the whole field, eating most of the pests. You can still use tractors in the field, because they can easily drive straight over the long grass.

Question: What's the strangest pest control you have ever heard of?

Answer: Definitely getting rid of mites on bees. Honey bees get tiny mites living on them, that kill them if there are too many. A company in Finland has found a way to get rid of the mites without using pesticides. They put a mirror just inside the beehive, so that bees flying into the hive get confused and fly straight into the mirror. As the bees bump against the mirror, the shock shakes off the mites, which fall into a tray underneath that the beekeeper can empty from time to time!

Qs

- Give two reasons why greenfly multiply very quickly on rose bushes.

- What did scientists think the *Euglandina* snail would eat? What did it eat?

- Why does a beetle bank increase the number of beetles in a field?

- Suggest a possible reason why the honey bee mites stay in the tray instead of going back onto the bees.

Activities

- *Discuss:* Do you think it is acceptable to introduce predators from foreign countries to control pests, if the right tests are done first? What would the right tests be?

- *Role-play:* Imagine you are in charge of buying the vegetables to sell in a supermarket. Listen to the arguments of an organic farmer and a farmer who uses pesticides, when they tell you why their produce is best. Think about what your customers will want, then decide whose produce you are going to buy. Explain why you made this choice.

Exploding pests!

Setting the scene

Throughout history there are many accounts of plagues of animal pests, such as the 'Egyptian Plagues' from the Bible, or the rats from the fairy tale *The Pied Piper of Hamlin.* You may well know of others. So what is it that controls the number of any species in an environment, and can plagues of different species really happen?

Modern plagues

Huge plagues of some species, such as grasshoppers, moths, ladybirds, locusts and even frogs, mice or crabs are not unusual. Scientists don't usually call them plagues though, because a plague is really a disease. Instead scientists talk about a 'population explosion'. These eyewitness accounts give you some idea of what a 'population explosion' is like.

Canada, 1990: 'We had to stop manufacturing chipboard because we couldn't sweep the moths away from the machinery fast enough. No one will buy chipboard with squashed moths in it.'

South Australia, July 1993: 'In the worst-hit areas there are about 175 mice per square metre.'

Ontario, Canada, October 2001: 'At first I thought it was snowing, then I realised it was ladybirds.'

Kansas, USA, June 2002: 'The lawn was literally a carpet of moving grasshoppers.'

What causes population explosions?

The population of any species depends on how much food there is, and how many predators there are. If the food increases or environmental conditions become more suitable, the population will go up, and if the number of predators increases, the population will go down. For example, a warm winter might make the population of an insect increase the following spring, because less were killed during hibernation. If good conditions make the population of a species increase very quickly, the number of predators may not be able to increase fast enough to 'keep up' and then there will be a population explosion.

Natural population explosions

Some species have natural population explosions every few years. One example is the gypsy moth, common in north-east America. Birds, rodents and beetles eat gypsy moth caterpillars, but in good conditions each female moth can lay up to 1500 eggs at a time, so the predators cannot eat the caterpillars as fast as new eggs are hatching. The population explodes. Then, just as suddenly, nearly all the moths die and the population falls back down to its normal numbers. Why? The answer is a virus that kills gypsy moths but no other species. As the number of gypsy moths goes up it gets easier for the virus to pass from moth to moth, so nearly all the moths catch the virus and die.

What if there aren't any predators?

Population explosions often happen if a species is introduced to an area where it has no predators. In the 1990s population explosions of tree frogs chirping all night began to keep residents in Hawaii awake. The frogs had probably arrived in shipments of garden plants, and had bred very quickly because none of the local predators ate them. It was estimated that there were more than 8000 frogs per hectare eating up to 46 000 insects per hectare and threatening native birds with starvation. In 2003 the situation became worse when another introduced species, the brown tree snake, began eating the frogs. Numbers of brown tree snakes increased rapidly, and brown tree snakes also ate the native birds! One scientist said 'We are studying the effects of giving the frogs pure caffeine to kill them, because the caffeine seems to be harmless to humans, native plants and wildlife.'

This tiny frog has a chirp as loud as a lawnmower, and it chirps all night!

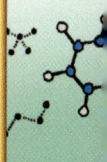

Qs

○ Describe two things that can cause the population of a species to go up.

○ Why don't predators stop a population explosion from happening?

○ Why does a population of gypsy moths have to increase to 'population explosion' levels before the moths are killed by the virus?

○ Describe two ways, one direct and one indirect, in which a population explosion of frogs on Hawaii harms the native birds.

Activities

■ Imagine you suffered a population explosion of a local species of insect or other animal. Discuss which species you would find worst, and what you would do about it.

■ Imagine you are a local pest controller in an area that has a population explosion of gypsy moths. Some local residents are demanding that you spray the moths with pesticides to kill them. Role-play the discussion you will have with them to explain the advantages and disadvantages of using pesticides to control the moths.

Diamonds are the industry's best friend

You know that there are different types of atoms, and they join together in different ways to make all the different materials we see. For example, each tiny particle, or molecule, of water is made from two atoms of hydrogen joined to one atom of oxygen. Carbon atoms are very good at joining to other atoms – there are even three different materials made from just carbon atoms!

Setting the scene

Question: *I've read that diamonds are made from carbon, but they can't be, can they? I thought carbon was a black, powdery stuff.*

Answer: Carbon is usually a black powder. It is the soot that is left when things have burned. But there are other types of carbon too. You can imagine the carbon atoms like a child's toy bricks, which can be put together in different ways to make different things. If the carbon atoms are all jumbled up, you get soot. If the carbon atoms are in round plates, like tiny Frisbees, you get graphite, which is the 'lead' you get in lead pencils. If the carbon atoms are stacked neatly, like a pile of bricks the builder hasn't started using yet, they make diamond.

Question: *What makes diamonds so special?*

Answer: Most people think diamonds are special because they are rare, they look nice and they can be cut so they reflect the light and sparkle. People in industry think diamonds are special because they are very hard. Diamond is the hardest material known; it is so hard it can be used to cut almost anything, even metal and stone.

Diamonds can be split along the edges of the crystal structure to make precious jewellery.

Question: *Where do diamonds come from?*

Answer: Diamonds are formed by high temperatures and high pressures deep in the Earth. When volcanoes erupt some diamonds are brought to the surface. Less than 20 tonnes of natural diamonds are mined each year in the whole world.

Question: *What are artificial diamonds?*

Answer: They are diamonds made in a laboratory. In the 1950s scientists worked out how to change graphite into diamond by making it very hot and squashing it. The 'industrial diamonds' made like this are tiny, like grains of sand, but they are very valuable in industry. They are mixed with metal to make saws or grinding tools, or drills for drilling through rock to find oil.

Question: *Can you make jewellery from artificial diamonds?*

Answer: You couldn't, but in the 1990s scientists worked out how to condense diamond crystals from gases with lots of carbon in them. Now they can make diamond crystals bigger than any diamonds found naturally. It is impossible to tell the difference between these artificial diamonds and natural diamonds just by looking at them. Natural diamonds usually have more impurities, chemicals other than carbon, in them than artificial diamonds, but there is no single test that can tell if a diamond was mined or 'grown' in a laboratory.

Qs

- Name three materials that are made from pure carbon.

- Why do people in industry think diamond is a special material?

- Suggest a reason why natural diamonds are very expensive.

- How is making large diamond crystals, suitable for making jewellery, different from the original way of making industrial diamonds?

Activities

- *Discuss:* Do you think the price of a diamond from a mine and the price of a diamond grown in a laboratory should be the same or different? What are your reasons?

- *Discuss:* In some areas of South Africa whole communities are dependent on the income from diamond mines. How do you think they feel about artificial diamonds?

- Imagine you are a jeweller selling jewellery made from artificial diamonds. Design an advertisement to convince your customers that your diamonds are as good as the 'real' diamonds from a diamond mine. It could be a poster or a radio or television advertisement.

Don't blink – or you'll miss it!

Setting the scene

You have learned that an element is a chemical made from only one type of atom. Most of the materials around us are not elements because they are made from several different sorts of atoms joined together. For example, water is hydrogen and oxygen atoms joined together. So how can scientists tell if something is an element, how many elements there are, and can they make them? Read these questions written to a science magazine.

Question: How many different elements are there?

Answer: There are 92 natural elements – elements that occur by themselves somewhere in the world. The periodic table is a list of all these elements, arranged in groups of elements that are similar to each other. There are about another 20 elements that scientists have managed to make in the laboratory, but these don't usually last very long – they change into different elements all by themselves!

Question: We've done lots of experiments at school to mix things together and to heat things, but I still don't see how scientists can tell whether something is an element or a compound. Can you explain?

Answer: Scientists found the natural elements using very accurate weighing, much more accurate than you can do at school. It's really to do with how atoms join together. Imagine you weigh a dish with an element in it, and then you do a chemical test such as heating, and weigh the dish again. If any other atoms have joined onto the atoms of your element, the dish will be heavier than it was. Elements never get lighter, whatever chemical tests you do to them. So chemists try lots of chemical tests, and if something never gets any lighter, they know it must be an element.

How many chemical tests do you know?

Question: I heard that scientists can make elements. Is that true?

Answer: Yes. Different elements have atoms of different weights. In the first half of the twentieth century, scientists found that by making light atoms travel very fast and firing them into each other, they could make them stick together to make one heavier atom. In the 1940s, scientists made plutonium like this, and since then they have made many other elements with strange names like technetium, americium, curium, einsteinium, californium, mendelevium, nobelium and lawrencium. Often the new elements are named after places or people. New atoms made like this don't usually last very long – they fall apart again into smaller atoms, but not always the same small atoms as you started with!

Question: How easy is it to tell if you have made a new element?

Answer: Not very easy at all. In 1999 scientists from Berkeley Laboratories in America announced that they had made the heaviest element ever, by sticking lead and krypton together, and they called it ununoctium. Atoms of ununoctium only lasted less than one thousandth of a second before they broke up. Two years later, in 2001, the same American scientists announced that their 'discovery' had been a mistake. They had spent two years repeating their experiments and were unable to get the same results again. So they decided that probably they had made a mistake in their first readings and ununoctium was never there at all!

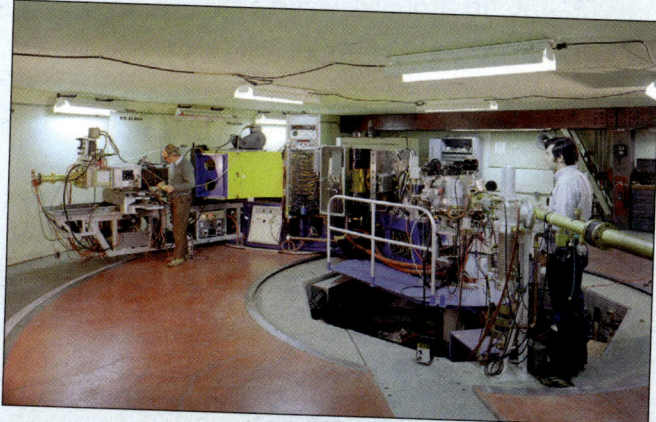

Qs

- What is an element?

- When scientists do chemical tests, what measurements can they take to find out if something is an element or not?

- List two elements made artificially that are named after places and three that are named after people.

- How did scientists at Berkeley Laboratories make ununoctium?

- Why did they decide they had made a mistake?

Activities

- The Berkeley scientists made a mistake. Discuss whether they were good scientists or bad scientists. Think about how they worked and what they did with their results.

- Draw a diagram to show how new elements are made by firing small atoms at each other. Your diagram should also show what happens to the new atoms shortly after they are made.

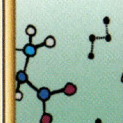

When will the Sun go out?

You know that there are less than 100 different types of natural elements, listed in the periodic table. But where did the different atoms come from, and do atoms always stay exactly the same? Astronomer Joji Swensen tells us more.

Setting the scene

Interviewer Hello, Joji. I understand you are going to tell us how scientists worked out what the Sun is made from and how old it is.

Joji That's right. It all began with the early Victorians who thought the Sun was a big ball of burning coal. That worked fine until they began to ask 'I wonder how long it will keep burning for?' They used Newton's work on gravity to calculate how heavy it was, and they found that even if it was all coal, it would only burn for about 1500 years! That was a real problem because even the early Victorians knew from the Bible that the Sun had been shining down on Earth for at least 6000 years! They couldn't find any chemical that would burn for that long.

Interviewer So what was it made from?

Joji Nobody knew, and it got worse because soon scientists like Charles Darwin showed that the Earth wasn't just thousands of years old, it was hundreds or even thousands of millions of years old, and the Sun had been there for all of that time too!

Interviewer So what happened?

Joji Well, in the 1890s there was a very important discovery – radioactivity. Scientists found atoms that changed into different types of atoms all by themselves, giving out energy and particles that they called 'radioactivity'. Radioactive changes give out much more energy than chemical changes do. If the Sun's energy came from radioactive changes, it could have been there for billions of years and still last for billions of years more.

Interviewer So the problem was solved?

Joji   Not quite. The scientists worked out that all the stars are like giant chemical factories, using radioactive changes to make all the elements we know from hydrogen gas. But the elements are made in a set order, and scientists can work out how long it takes to make each element. By looking at the colour of the light from the Sun, the scientists could find out how much hydrogen was left in the Sun, so they could calculate its age. They found the Sun is about 4.7 billion years old – which just isn't old enough to have made all the different elements in the solar system.

Interviewer   So where did they come from then?

Joji   They must have been there when the solar system first formed. That means the Sun and all the planets must have been made from a cloud of gas that already had lots of different elements in it. And the only way that could happen was if the cloud came from an earlier star that had just exploded.

Interviewer   You're kidding!

Joji   No, I'm not. We now know that our Sun is a 'second-generation star' made from the bits left behind when an older star exploded. Five billion years ago our Sun didn't exist, but another star would have been there in its place. Perhaps it even had planets of its own going round it just like we go round the Sun today.

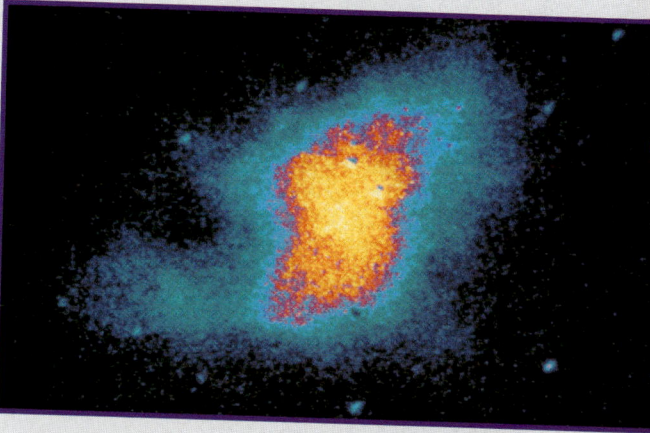

Our whole solar system formed from the exploded remains of an old star.

Interviewer   That's amazing.

Qs

○ Explain how the Victorians worked out that the Sun is not made from coal.

○ What is radioactivity?

○ How long could the Sun 'burn' for if its energy came from radioactive changes?

○ What is a 'second-generation star'?

○ What told scientists that the Sun must be a 'second-generation star'?

Activities

■ Since radioactivity was discovered in the 1890s, we have found out much more about it. Discuss radioactivity in your group – you will probably be surprised by how much you know. Make a diagram or concept map to show what you know and share it with another group.

■ Make a timeline that shows the major discoveries mentioned by Joji Swensen, in the correct order. Do some research to find dates for one or more of the discoveries.

The philosopher's stone

Most of the materials around us are made from different types of atoms joined together. You probably know lots of chemical reactions to join atoms together in different ways, or to make them separate again. But to the early experimenters it must have seemed like magic to warm or stir materials and have completely new materials appear, almost from nowhere. So what experiments did they try to do, and what did they learn?

Many of the alchemists of the Middle Ages were monks.

Magic or science?

We get our words 'alchemy' and 'chemistry' from the ancient Greek words 'al chemeia', meaning 'the Egyptian Art', because the Greeks learned a lot about doing chemical experiments from the ancient Egyptians. Most people thought alchemists were magicians, and many alchemists did do 'magic' and make prophesies, as well as doing experiments. One alchemist's prophesies included flying machines, submarines and high-speed horseless carriages! Although it must have seemed like magic to make new materials appear, almost from nowhere, the early alchemists actually invented a lot of the methods that modern chemists still use today.

The philosopher's stone?

For over 3000 years, from about 2000 BC to about 1650 AD, alchemists were searching for two things. They wanted to make an 'elixir of life' that would give long life and good health to anybody who drank it. And they wanted to find a 'philosopher's stone' that would turn all other metals into gold. Several early alchemists were said to have found the philosopher's stone, but they didn't share the secret with anyone! One of them was Nicholas Flamel, who died in 1418 at the age of 88, and he certainly died a very rich man.

Mad alchemists?

Today we use heating and condensing to make lots of pure substances, but the alchemists discovered it first! They used it to make pure sulphur and pure mercury. Most volcanic rock contains sulphur, which turns to a gas when the rock is heated, and condenses as a yellow solid on cold surfaces nearby. Imagine their excitement when they heated mercury compounds and found mercury, a silvery metal, condensing on surfaces near by and running down! They called mercury 'quicksilver'. Unfortunately, mercury gas is very poisonous, and it sends people mad, so perhaps people were right when they thought alchemists were mad.

Drunken alchemists?

Alchemists also invented fractional crystallisation of sea water. Sea water has common salt in it, but it has lots of other chemicals in it as well. When sea water is evaporated slowly, crystals of common salt settle at the bottom first, then crystals of other chemicals. We still use this way today to make table salt. In about the 1200s, alchemists also learned that they could increase the amount of alcohol in alcoholic drinks by distilling them. It was about then that lots of countries introduced laws to control drunkenness! We still use distillation today to make petrol from crude oil.

Making gold

To us it sounds stupid to think you can make gold from other metals, but it made sense to the early alchemists. Most metal ores look metallic. When they are heated, pure metal runs out, so it seems as if one metal has changed to another metal.

These are some of the metal ores that alchemists thought were real metals. Can you see why?

- Who were probably the first people to do chemical experiments?

- What two things were alchemists searching for?

- Name two substances that alchemists found by heating and then condensing.

- Give two examples of things alchemists learned how to do, that we still use today.

- Alchemists made many discoveries about different materials and the ways they react with each other. Discuss some of the ways of working, or characteristics, that you think would make a good alchemist.

- Choose one of the things alchemists invented or discovered. Write a 'newspaper report of the day' to tell people about it.

All mixed up!

You have learned the names and properties of several different metals. You know that we can use different metals for different things, because they have different properties. For example, we make aeroplanes from aluminium because aluminium is light, and cranes from steel because steel is strong. But what can we do if we can't find a metal with just the right properties? Read this science magazine to find out.

Setting the scene

Alloys – why bother?

An alloy is made by mixing two or more metals together, because the alloy has more useful properties than the separate metals. People have known about alloys for thousands of years; they made bronze by adding tin to copper to make it stronger and more flexible; they made steel by adding carbon to iron to make it stronger. Modern scientists are experimenting with lots of different mixtures of metals to make new alloys. The new alloys are usually expensive, but they have some very interesting properties.

Editor's note: Some people say steel isn't a true alloy because carbon isn't a metal. Look through the other alloys on this page – are there any others that are not true alloys?

Liquidmetal – better than metal and plastic

In 1992, researchers at the Californian Institute of Technology made a new alloy, called Liquidmetal, by mixing together five metals, titanium, copper, nickel, zirconium and beryllium. The atoms in most metals are arranged in a crystal pattern, which means sometimes they crack along the joins between the crystals. The metals in Liquidmetal have different size atoms that won't fit together neatly, so when the metal cools down the atoms stay all jumbled up instead of making crystals. This makes Liquidmetal very strong, more than twice as strong as steel, but it can be moulded like plastic.

Andre Agassi backs Liquidmetal!

The top tennis star, Andre Agassi, chooses a racquet made with Liquidmetal. The Liquidmetal is very strong, and it is bouncy compared with other metals. So when Andre hits the ball, nearly all of his energy is transferred to the ball instead of some of it being absorbed by the racquet. This means Andre can hit the ball further and faster than he could with his old racquet.

Liquidmetal takes off!

Liquidmetal has been on four space shuttle missions already – with more planned. A NASA spokesman told us, 'We are really excited by Liquidmetal's properties and think it probably has a great future in space. It is much stronger than steel and it lasts much, much longer without wearing out. But it is still at the experimental stage at the moment – we have got to check out how it works in space before we make anything vital from Liquidmetal. We just can't take risks with astronauts' lives.'

Gum metal – like really tough chewing gum!

Gum metal is the newest 'wonder metal'. Japanese researchers made it from a mixture of titanium, niobium, tantalum, zirconium and oxygen. They called it 'gum metal' because it is bendy and stretchy, like chewing gum. It can be stretched to twice its original length and it just springs back. It has been used to make spectacle frames that can be bent and twisted almost to right angles, and just spring back to their original shape! Spacecraft designers say it may be useful in space because it keeps its springiness at temperatures from $-200\,^{\circ}\mathrm{C}$ right up to $200\,^{\circ}\mathrm{C}$.

Qs

- What is an alloy?

- Why do some people say steel is not a true alloy?

- How is the arrangement of atoms in Liquidmetal different from the arrangement of atoms in pure metals?

- Explain in your own words why Andre Agassi chooses a Liquidmetal racquet.

Activities

- The manufacturers of Liquidmetal used it to make sports equipment before anything else. Discuss the reasons why sports equipment is a good choice to test out a new metal.

- Design something that could be made using Liquidmetal or gum metal. Explain what makes Liquidmetal or gum metal useful for your design.

Catching the drug barons

You know that there are different types of atoms and they join together in different ways to make all the materials we see around us. Carbon atoms are very important because they form the compounds that make up all living things. But there are actually two slightly different types of carbon atoms, with slightly different weights. Scientists have found a way to use these to catch the people growing the plants which are made into illegal drugs. Read this customs officer's diary to find out more.

HEATHROW, DECEMBER 2001

I've been looking back over our figures for the drugs we seized last year. Even when you count the number of people arrested by the police for supplying illegal drugs, I still feel we are not really winning this war. It would help a lot if we could catch some of the growers. It's easy to find out exactly where the drugs were grown – scientists look at pollen grains trapped in the supply of drugs. The exact mix of pollen grains from different plants gives biologists a very good idea of the exact area where the plants were grown, but that doesn't really help. The drug barons tend to store the illegal drugs until they think it is safe to bring them into this country. We can't tell whether the drugs come from plants grown this year, or ten years ago. The drug barons keep moving their fields of plants around, so if we go to where we know the drugs were grown, the grower will probably be growing coffee there instead.

Finding illegal drugs is only part of solving the problem.

HEATHROW, DECEMBER 2003

The year is coming to an end and I feel much more cheerful about our chances of eventually getting rid of illegal drugs. I think it is getting harder and harder all the time for people to grow and supply illegal drugs without getting caught. There was a real breakthrough earlier this year, when some scientists from Australia found a way to tell exactly when a drug crop was growing – it's all to do with different types of carbon atoms.

Apparently there are two different types of carbon atoms, normal carbon – called carbon-12, and radioactive carbon – called carbon-14. Nearly all of the carbon is carbon-12, but the proportion of carbon-14 keeps changing. The nuclear weapons tests of the 1960s, 1970s and 1980s increased the proportion of carbon-14 in the atmosphere, but this proportion has been going down steadily ever since because most of the weapon testing has stopped. Growing plants absorb carbon dioxide from the air, so the proportion of carbon-14 in a growing plant is exactly the same as the proportion of carbon-14 in the atmosphere. When the plant stops growing, it stops absorbing carbon dioxide, so the proportion of carbon-14 in the plant stays the same as it was when the plant stopped growing.

Scientists can measure the proportion of carbon-14 in a sample of illegal drugs and tell whether that drug came from plants growing this year, or from plants that grew ages ago. So we then know whether it is worth going to the place where the plants grew, to find the grower.

Finding illegal drug plants in a dense rainforest is very hard, but now tests on the drugs can tell scientists exactly where and when to look.

Qs

- How can scientists tell where drugs plants grew?

- Why is this information not enough to catch the growers?

- Explain how the proportion of carbon-14 in a sample of drugs tells us when the drug was grown.

- Do you think the proportion of carbon-14 in drugs will depend on where the plants were grown, as well as when they were grown? Give your reasons.

Activities

- The UK spends millions of pounds trying to catch people supplying illegal drugs. Discuss whether or not you think this is money well spent.

- Teenagers are often thought to be those most likely to be targeted by drug suppliers. Design a poster or advertisement warning young teenagers of the dangers of drugs.

43

Strange weathering

Rocks don't stay the same for ever. Changes in temperature, rain, water and wind, gradually break up and wear away rocks on the Earth's surface. We usually imagine that, as this weathering happens, the rocks just gradually get smaller and smaller, but actually it's much, much stranger than that. Geologist, Gill, is going to tell us more.

Interviewer Good morning, Gill. The way rocks weather can be very strange. Is that right?

Gill Yes, but before I go on, how much do your listeners know already?

Interviewer Quite a lot. They know that cold weather can freeze the water in rocks and make bits break off. They know that rainwater reacts chemically with some rocks to make them dissolve and wash away.

Gill Excellent. That's a great help, because I want to talk about how water can make some very strange shaped rocks.

Interviewer Well, start with something simple, please.

Gill Alright, we'll start with sea cliffs. What happens when the sea crashes against a rocky cliff?

Interviewer I don't know. It splashes everywhere?

Gill Yes, but the rock wears away, just like when rain falls on it or a river runs over it. But only the bits hit by the sea wear away, so on tall cliffs you get the bit at the bottom wearing away and the bit at the top staying there. We call it an overhanging cliff.

Interviewer Is that it?

Gill Not quite. Sometimes there are bits of hard rock and soft rock in the cliff. Then the soft rock wears away and you get pillars or arches of rock left. You can see examples of this at The Needles on the Isle of Wight or at The Green Bridge of Wales in Pembrokeshire.

Interviewer So do these interesting shapes only happen by the sea?

Gill No, they can happen anywhere where water flows over rocks. In Derbyshire and Yorkshire water flowing though limestone rock underground has dissolved so much rock that there are huge caves. Lots of them are open to the public – you can even go for an underground boat trip though Speedwell Cavern in Derbyshire.

Interviewer Are there any strange rock shapes on the surface?

Gill I think one of the strangest is limestone pavement.

Interviewer That sounds very dull!

Gill I don't think it is. The most famous limestone pavement is at Malham Cove in North Yorkshire. It is like a pavement made of flat topped blocks of rock, called clints, with huge cracks, called grykes, between them. The cracks go down further than you can see, and you have to be careful not to drop things in them or get your foot stuck in them. Over thousands of years, water has flowed through tiny cracks in the rock, dissolving it away and leaving the huge cracks.

Interviewer I once saw small round holes in a rock. It looked like someone had been drilling holes in it. What causes that?

Gill Again it's water dissolving part of the rock. The holes fill with water, which makes the dissolving happen faster. I once found a small rock with a hole right through it, like a bead. People used to keep rocks with holes in them to bring them good luck, because they thought they were magical.

Interviewer Amazing! Thank you for talking to us, Gill.

Qs

- Describe briefly two things that can cause weathering of rocks.

- What causes overhanging sea cliffs?

- What is formed when water dissolves rock underground?

- It is extremely rare to find underground caves in granite rock. Suggest one thing this tells you about the granite.

Activities

- Acid rain is rainwater that is more acidic than normal. Discuss what effect acid rain might have on the weathering of rocks. What sort of rocks will be affected most?

- Limestone pavement is quite rare, and it has several species of rare and unusual plants growing in the grykes. But it is also very popular as a stone for garden rockeries. Role-play a disagreement between a company that wants to dig up limestone pavement for rockery stone and a conservation group who wants to protect it. Try to make your side of the argument as convincing as possible.

Is bottled better?

You know that rainwater can react chemically with the rocks it falls on, and this can cause them to weather and fall apart. But what happens to the rock? Is it left in the water? Does it affect the water we drink? And is there any difference between bottled water and tap water?

Setting the scene

QUICKMART Sparkling Mineral Water

Enjoy the taste of mineral water at a price you can afford. Just relax and forget all those worries about chemicals in tap water. Our mineral water is totally pure – it is bottled straight from the spring where it bubbles from the ground. We can guarantee that once you have tasted our delicately sparkling mineral water you will never want to drink tap water again!

Tideswell Spring Water

Tideswell Spring Water has a unique, natural flavour that comes straight from the hills! You can taste the clean, fresh, flavour of Derbyshire's Peak District in every glass! Every drop of our Spring Water has been gently purified by the natural power of thousands of metres of limestone rock!

Editor's comment: Although today many people claim mineral water tastes better than tap water, the Victorians thought differently. They thought mineral water tasted disgusting, but they drank it anyway because they thought it was good for them! Mineral water and tap water come from different sources. Tap water is water that has collected above ground, in rivers or reservoirs, and then been filtered and disinfected, to remove particles of dirt and plants and to kill bacteria. Chlorine is added to keep the water pure in the pipes, on its way to our taps. Some people claim they can taste the chlorine in tap water, other people can't. Mineral water is water that has seeped through underground rocks, then bubbled up out of springs. It does not need purifying because it has been filtered by the rock. It does not naturally 'sparkle' – 'sparkling' mineral water is made by dissolving carbon dioxide gas in the 'still' spring water.

SPRING FRESH mineral water

Drink SpringFresh as part of a healthy diet! SpringFresh is rich in the natural minerals and trace elements that we need for healthy growth and a tough, modern lifestyle. Or try SpringFresh Scottish for a stronger, more rugged set of minerals, ideal for the more active, sporty individual!

Editor's comment: Don't believe every advertisement you read! There's no such thing as 'stronger, more rugged minerals', but mineral waters from different places will have slightly different flavours, because they have seeped through different types of rock, dissolving a slightly different mixture of minerals. If mineral water drips very slowly from the roof of a cave, for example, the minerals settle out, and stalagmites form where the water drips. The stalagmites are slightly different colours in different caves, because the mix of minerals is different. Explorers in Heathery Burn Cave in the north of England once found the skeletons of a Bronze Age family and, nearby, a hoard of treasure hidden under a layer of stalagmite, making them wonder if Bronze Age people placed treasures under dripping water on purpose, to preserve them.

- People often claim that both tap water and mineral water have a flavour. Where does the flavour of each come from?

- Why is tap water purified?

- Why doesn't mineral water need to be purified?

- How is mineral water made 'sparkling'?

- Unopened bottles of mineral water can be kept for months but, once opened, they should be kept in the fridge and drunk within a couple of days. Why do you think this is? What precautions do you think the bottlers of mineral water take so that the water can be kept for so long when unopened?

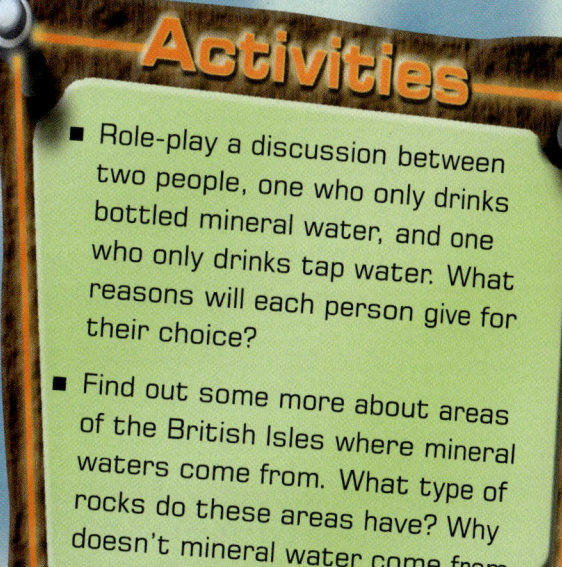

Activities

- Role-play a discussion between two people, one who only drinks bottled mineral water, and one who only drinks tap water. What reasons will each person give for their choice?

- Find out some more about areas of the British Isles where mineral waters come from. What type of rocks do these areas have? Why doesn't mineral water come from other places?

The riddle of the Sphinx

You have probably heard of the Great Sphinx of Egypt. It is an ancient carving, over 70 metres long and 20 metres high, of a lion's body with a man's head, which crouches near the Great Pyramids of Giza, facing east towards the sunrise. Many people claim that weathering patterns on the Sphinx can tell us how old it is, but they still can't agree an age. So what do they say?

Setting the scene

The Sphinx is certainly impressive. People who have studied it agree that it was not made from blocks of rock, but was carved from the solid rock floor of a man-made quarry. The body is carved from much softer limestone than the head, and shows much more erosion. People think that some of the damage to the face was done by Napoleon's artillerymen using the Sphinx for target practice! Read these imaginary letters discussing how old the Sphinx is.

Dear Dr Schoch,

The Sphinx was clearly built at the same time as the great pyramids next to it, that is in about 2500 BC. We know that the style of head-dress shown in the carving was very common in Egypt at the time the pyramids were built.

Dear Mark,

I don't think it was built at the same time. The king who built the great pyramids boasted about all the other things he built, but there aren't any records at all about building the Sphinx. That's odd. Also the pyramids show horizontal marks where they have been weathered by sand and wind, but the erosion on the Sphinx is different. The marks are vertical, just like the erosion made by heavy rainfall. Egypt had a much wetter climate between 7000 BC and 5000 BC. I think the Sphinx was built then, or even earlier.

Dear Dr Schoch,

I agree the Sphinx doesn't show sand and wind weathering, but it was buried up to its neck in sand until the nineteenth century. At night it gets very cold and condensation settles on the sand and soaks in. Just a few centimetres below the surface, the sand can be wet through, even though it's dry on the surface. The erosion on the Sphinx's body was probably caused by this wet sand, not by rainfall at all.

Dear Mark,

I still think the erosion was caused by rain. The rocks underneath the Sphinx contain channels and holes exactly like those caused by flowing water in other places. I think the Sphinx was built even earlier than the wet period in 7000 BC to 5000 BC, long before the beginnings of the Egyptian civilisation that we know about, before records began. I'm fairly sure it was built in about 10 500 BC, because of the way it faces. Over thousands of years, the Earth's orbit wobbles and different constellations of stars appear on the horizon at sunrise. The Sphinx is a lion, and if it were built in 10 500 BC then at sunrise it would be looking straight at the constellation of Leo, which is also a lion. This can't be just coincidence.

Dear Dr Schoch,

I think it is just coincidence. We don't even know if there were any people in Egypt that long ago and, even if there were, they certainly wouldn't carve a head in the style of Egyptians of 2500 BC.

Dear Mark,

I think the king who built the pyramids re-carved a new head on the Sphinx that was already there, to change a lion's head into a head that looked like him.

Qs

- What is the visible difference between weathering caused by sand and wind and weathering caused by rainfall?

- If the erosion on the Sphinx were caused by rainfall, what would this tell us about the age of the Sphinx?

- Mark tells Dr Schoch that the Sphinx is the same age as the Pyramids because of the style of the carving. Is this evidence or opinion? List some other examples of evidence and opinion from the article.

Activities

- Discuss how old you think the Sphinx is. Whose arguments do you find most convincing, Dr Schoch's or Mark's? Do you think we will ever find out for certain how old the Sphinx is?

- Use the internet to find some more pictures of the Sphinx. Make a Powerpoint presentation about it. Include information about how old it is, if you wish.

As clear as glass

Setting the scene

Rocks are formed in different ways, in many, many different types. They can be soft and powdery, like chalk or talcum powder, or gritty and easy to wear away, like sandstone. Clear crystals, like rubies or emeralds, are formed when molten rock cools down slowly. But if the molten rock cools quickly, it makes glass – so glass is really a sort of rock! But it is a rock that is only found in odd places.

Obsidian – evidence for early man?

SCIENTISTS SAY OBSIDIAN FOUND NEAR A VOLCANO SHOWS EARLY MAN WAS THERE! Obsidian is a hard, black natural glass often formed in volcanoes. 'If the obsidian is chipped or flaked, it gives very sharp edges – better than a lot of modern knives,' one scientist told us. 'Early man used to make arrow heads, spear heads, knives and razors from obsidian. The earliest obsidian tools found so far were made in about 75 000 BC. We don't think ours are quite that old.'

Desert meteorites. 'Are we safe?' ask locals.

EVIDENCE of meteorite strikes have been found in the Sahara Desert! Local communities ask 'Are we safe?' Scientists have found pieces of glass in the sand. They say the glass was made by a large meteorite melting the sand and turning it to glass. 'These things always seem to happen here,' one worried local complained, 'life here is hard enough without meteorites falling on our heads.'

We asked a scientist to explain. 'Local people don't need to worry,' he told us. 'Large meteorite strikes are very, very rare, and this one may be thousands of years old. They happen all over the world; it's just that we can see the evidence in the desert because it doesn't get covered up by soil and plants.'

'We're not certain yet it was a meteorite,' another scientist told us. 'We have to look carefully at the shape of the glass,' she explained. 'Meteorite strikes leave "splashes" of glass, called tektites, around the edge of a crater. Lightning leaves long tubes of glass, called fulgurites, where the lightning went through the sand.'

Glass to make former nuclear test sites safe?

SCIENTISTS ARE USING GLASS to try to make former nuclear test sites safe. The first nuclear bomb was tested at the Trinity Test Site in the New Mexico Desert. The centre of the explosion got hotter than the surface of the Sun, and the sand melted, turning into a light green glass. Scientists called the glass 'Trinitite', after the name of the test site. Scientists have found that the glass traps some of the radioactive particles, sealing them in and making them harmless for thousands of years. So now they are trying to make other nuclear test sites safe this way. They use electricity to heat up the sand until it melts, turning to glass and trapping the radioactive uranium or plutonium and making it safe.

Qs

- What made obsidian, or volcanic glass, very good for making prehistoric tools?

- List three ways in which natural glass can be made.

- Most nuclear test sites are in deserts. Can you suggest why?

- Do you think glass formed in a nuclear explosion would be shaped more like glass from a meteor strike, or more like glass from a lightning strike? Suggest a reason why.

Activities

- Do you think testing of nuclear weapons is necessary? Discuss your ideas.

- Find out some more about the dangers of radioactivity. Design a poster to warn local children, who may not be able to read, to stay away from radioactive areas.

It's the Earth's fault!

You know how rocks are worn away and new rocks made in the rock cycle. Most of the time the rock cycle is a very slow process, but it can be dramatic and violent. Then we see reports on our television news of serious earthquakes, with people killed and buildings destroyed. So why do earthquakes happen, and can we tell when? Our interviewer talks to an expert, Ruth, to find out more.

Setting the scene

Interviewer Today we have Ruth on our programme, who is an expert on earthquakes. First, tell us why we get earthquakes, please Ruth.

Ruth That's a very big question! The Earth feels solid, but it isn't really, it's a big ball of gloopy, molten rock, with a thin layer of solid rock on the outside. The solid layer is made up of lots of flat plates of rock that slide around very, very slowly on the molten rock. In some places, called 'faults', the edges of the flat plates are moving further apart, or sliding up over each other.

Interviewer I think I've got that, but I don't see why that causes earthquakes.

Ruth It's because the plates don't slide smoothly. They move jerkily because they keep sticking on each other. An earthquake happens when two plates that are sticking, suddenly jerk apart and move over each other a little bit, perhaps only a few millimetres. The more they move, the bigger the earthquake. In a really large earthquake they may move a metre or more.

Interviewer Can earthquakes happen anywhere?

Ruth No, only at 'faults' – the edges of the plates of rock, or cracks in the rock plates. Most of the big earthquakes happen at well known 'faults' round the edge of the Pacific Ocean.

Interviewer What do you call a big earthquake?

Ruth   The size of earthquakes is measured using the Richter Scale. People may not even feel earthquakes that are less than 5 on the Richter scale. Over 6 is a big earthquake that can seriously damage buildings and at over 7 most buildings collapse completely. There are about 18 earthquakes each year of magnitude 7 or bigger, but lots of them happen in areas away from buildings and people.

Interviewer   Can you tell how much damage an earthquake will cause?

Ruth   Not really. It depends what the buildings are like. We are still finding out how to make buildings that wobble but don't fall over, and we have found out that in areas likely to have earthquakes you should not build on some types of soil.

Interviewer   Why is that?

Ruth   Because some types of soil flow like liquids when they are shaken, so all the buildings get swept away. We call this liquefaction.

Interviewer   Can you predict when earthquakes are going to happen?

Ruth   I really wish we could – it would save so many lives. We are getting better at it, but earthquakes are all so different.

The arm on a seismograph machine swings backwards and forwards to show how much the Earth is shaking.

Sometimes there are several tiny earthquakes before a big one, so we can warn people – but then sometimes the big one doesn't happen! Sometimes there is no warning at all, like the 1976 Tangshan earthquake, which was magnitude 7.6 on the Richter Scale and killed 250 000 people.

Interviewer   Well, thank you for talking to us, and good luck with your research.

Qs

- What part of the Earth is solid rock?

- What scale is used to measure the size of earthquakes?

- Where, in the world, do most large earthquakes happen?

- Describe two things scientists have found out in their studies about how much damage earthquakes cause.

Activities

- Some areas of the world are well known for having large earthquakes. Discuss some of the reasons why people still live in these areas, even though they know earthquakes might happen.

- Find out some more about the types of buildings that survive earthquakes and those that fall down. Can you find out anything about new building methods which make buildings more likely to stay up in an earthquake?

Is smoking good for worms?

We know that soil is mostly ground up rock, but we still don't tend to think of the rock cycle as linked to living things, but it is. In the 1970s, eight years after man first walked on the Moon, scientists exploring the ocean floor discovered links between rocks and living things far stranger than anything in science fiction.

etting the scene

The rock cycle and living things

Whenever we burn coal, a type of rock, carbon is released into the atmosphere as carbon dioxide. The carbon dioxide is absorbed by plants, and may eventually turn back to coal again. Shellfish absorb calcium from their food, and use it to build shells. When they die the shells may become part of the sedimentary rock, limestone. Eventually the limestone dissolves in rainwater or sea water, and the minerals can be absorbed again by more living plants or animals.

What are black smokers?

In 1977, scientists in a deep-sea submersible called Alvin found tall chimneys of rock rising from the seabed and spitting clouds of black 'smoke'. They called them 'black smokers'. The black smokers occur near volcanoes, along mid-ocean ridges. Sea water seeps into cracks in the seabed, and reacts with hot magma welling up through the Earth's crust. Many chemicals, including metals and hydrogen sulphide, dissolve in the water. The mineral-rich sea water rises back to the surface through vents, where some minerals crystallise to build the tall chimney and others precipitate out to give the black 'smoke'. Black smokers can be huge. One, that scientists called Godzilla, is 13 stories high and 12 metres across.

Life at a black smoker

Scientists were amazed to find the black smokers teeming with animal life. They did not see how it could survive. The water temperature could be as high as 400°C, it was more acidic than vinegar and full of toxic chemicals that would kill most known living things. The pressure was huge, about the equivalent of you having 20 elephants standing on your head all the time. It was totally dark so there was no plant life and, apparently, no food. Yet here were vast communities of animals, including giant worms one and a half metres long and giant clams as big as dinner plates. What could they be eating?

Truth is stranger than fiction

All food chains need a source of energy, so scientists looked for the energy at the black smokers. They found that hydrogen sulphide reacts with oxygen in the sea water to give off energy. Huge colonies of bacteria were using this energy to change carbon dioxide in the water into sugars, just like plants use the energy of sunlight to change carbon dioxide into sugars. Some creatures were eating the bacteria but others, like the giant worms, remained a mystery because they had no mouths and no digestive systems. Scientists eventually found out that the worms act as a home for the bacteria in exchange for food. The worm absorbs all the ingredients the bacteria need from the sea water, a blood system carries the ingredients to a special organ packed full of the bacteria, and then carries some of the sugars made by the bacteria around to all other parts of the worm's body. Scientists are now wondering whether life on Earth began around black smokers, long before life forms dependent on plants evolved, and whether other planets might have similar life forms that use chemical energy instead of sunlight.

Qs

- What causes the black 'smoke' from black smokers?

- What type of energy is coming out from the vents, as well as chemical energy? How do you know?

- The giant worms and the bacteria have a 'symbiotic' relationship; they both gain something from each other. The giant worm gains food from the bacteria, but what does the bacteria gain?

Activities

- Scientists find it very hard to study the animal life from black smokers. Discuss some of the reasons why, and what the scientists might be able to do to overcome some of the problems.

- Work with a partner to draw an energy diagram, in any style you like, for a black smoker. Try to show where the energy comes from and how it is used.

Feeling hot, hot, hot!

Setting the scene

You are used to using a laboratory thermometer to measure the temperature of something, but have you ever wondered 'What happens if the liquid goes right out off the top of the thermometer?' or 'What happens if the liquid freezes?' These are things that other scientists have wondered about too, so they have invented other types of thermometers that can measure the temperature of all sorts of things, such as ovens or right inside your body.

Colour-change forehead thermometer

No more guessing whether or not your baby has a temperature. No more trying to hold a struggling toddler still while you measure her temperature with a glass thermometer under her arm. This thermometer is just a simple, plastic strip that you place on her forehead. Numbers written in different coloured dyes show up at different temperatures, so you can see at a glance how ill she is.

Note: Normal body temperature for humans is 37°C. Consult your doctor if your baby or toddler has a high temperature.

Ear thermometer

Is it time you replaced the old thermometers in your clinic or surgery? Our newest ear thermometer is even more accurate than ever before. Its new comfortable shape means the sensor rests one centimetre from the eardrum and measures the infra-red heat radiation coming from the eardrum. Tests have shown that it is much more accurate than under-the-tongue clinical thermometers or forehead thermometers, because it measures temperature deep in the body and it does not need to make a good contact to get an accurate reading.

This temperature sensor detects heat radiation from the eardrum. It is a very accurate way to measure body temperature.

Cooking thermometer

We are specialists in the catering trade. We know how important it is to prepare those 'microwave ready meals' at exactly the right temperature – just hot enough to cook them but not hot enough to destroy the flavour. Our Thermopile

Thermometer has a probe, made of two different types of wire joined together, that you can put right into the centre of the food. An electric current flows in the thermopile to show how hot the food is. The hotter the food, the bigger the current. The current is converted to a simple, digital, temperature reading that shows up on the wipe-clean display.

Radiation thermometer

Perfect for pottery kilns, metal smelting furnaces or even astronomy! As things get hotter, they glow red, then orange and finally white hot. Our radiation thermometer measures the colour of the radiation coming off to tell you just how hot your kiln or furnace is. No more china mugs cracking because the kiln was too hot! No more aeroplanes crashing because the metal wasn't made right! And our radiation thermometer works from a distance too; you don't even need to go near the hot object – astronomers can even use it to tell how hot stars are!

Magnetic thermometer

Fed up with all those thermometers that tell you how hot something is? Our thermometer tells you how cold something is! But it can only be used for really, really cold things, such as liquid nitrogen or even colder. It uses special chemicals that get more magnetic as they get colder. The stronger the magnetism, the colder it is.

- List two things you could measure the temperature of with an ordinary laboratory thermometer and two things it would be unsuitable for.

- Give two reasons why an ear thermometer is more accurate than a forehead thermometer.

- Why is it an advantage for a radiation thermometer to be able to measure temperature 'from a distance'?

Activities

- A cooking thermometer has a probe that can be put into the centre of the meat, or other food. Discuss why cooks measure the temperature here instead of on the surface of the food. Do you think the temperature really matters? Give a reason for your answer.

- One advertisement hints that aeroplanes would crash unless the manufacturers used a radiation thermometer! This isn't true, but can you find out about some things where temperature is really important? What would happen if the temperature were wrong?

Speedy suppers

Setting the scene

Conductors are materials that heat energy flows through easily, and insulators are hard for heat to get through. You know that saucepans are made from metal, to conduct heat into the food, and have plastic handles so the heat isn't conducted into your hand. But you probably don't use saucepans very much, you probably heat up lots of your meals in a microwave oven. So are microwave ovens really any different from, or better than, an 'ordinary' cooker?

'Manufacturer told me lies' claims unhappy shopper

MRS THOMAS, 59, is very unhappy with the microwave oven she bought for herself as a Christmas treat. She claims that the company who sold it to her told her lies. 'I only bought it because they told me it would cook my food much faster,' she told our reporter. 'I eat lots of pasta and rice dishes, and the microwave is absolutely useless. It's no quicker than the old cooker I've had for years – it might even be slower.'

Mr Upton, a spokesman for the microwave oven company, said 'We are very sorry Mrs Thomas is unhappy with her oven, but we didn't tell her lies. Microwave ovens do cook most foods much more quickly than ordinary cookers, because waves of microwave energy pass right into the middle of the food and are absorbed by the water in the food. We advise customers not to use the microwave oven for dried foods, because the waves of microwave energy pass straight through dried foods, so the food doesn't cook until it has soaked up water. You shouldn't use metal dishes in a microwave either, because microwaves can't get through the metal to the food. We will of course refund Mrs Thomas' money if she returns the oven to us.'

Young chef's TV disaster

Hayley Moss, our talented young local chef, came a very disappointing seventh in the Regional Finals. She told us what went wrong. 'It was the potatoes. Everything else went really well. For my 'supper dish' I decided to do small baked potatoes with a creamy tuna filling. I was a bit short of time, so I popped the potatoes in the microwave oven to bake, but I forgot to prick them. They exploded. It made a dreadful mess in the oven.' One of the judges added, 'Hayley was very unlucky – it was just the sort of mistake anyone could make. The microwaves turn the water inside the potatoes to steam. If you don't prick the skins, the steam can't get out and it makes the potatoes explode.'

MAN IN CANADA KILLED BY MICROWAVES

TELEPHONE COMPANIES OFTEN USE MICROWAVES TO TRANSMIT PHONE MESSAGES. Safety barriers prevent workers accidentally getting into the microwave beam. A watchman arriving for work on Boxing Day morning 1998 made a gruesome discovery. The night watchman had placed a deckchair in the microwave beam, which always felt slightly warm, and settled down to enjoy a few beers. The power of the microwave beam had been turned up to cope with all the people telephoning relatives on Christmas evening, and the night watchman had been cooked!

Editor's note: This story turned out to be a hoax, but an accident like this really could happen.

Qs

- Why aren't microwave ovens good at cooking dried food?

- Explain in your own words why Hayley's potatoes exploded.

- Suggest why the beam of microwaves at the telephone company felt warm.

- Microwave oven doors always have a metal grid inside the glass. This is not to make the glass stronger. Can you suggest what it might be for?

Activities

- Make an advertising leaflet for a microwave oven. Make sure you include information about what food you can cook in the oven and what equipment you should avoid using in the oven.

- Scientists still can't agree about whether or not microwaves can affect our health. Do some research to find out more about some of their concerns.

Baked Alaska

For years and years scientists argued about whether or not the Earth was getting warmer. Now scientists agree that global warming really is happening, but they still can't agree about how much the temperature will change, how quickly the changes are happening and whether the Earth will end up like an oven or like a snowball! So what makes it so difficult?

tting the scene

Do you think less ice is good news or bad news for penguins?

Question: Some scientists seem to think global warming will get better, others say it will get worse. Why don't they know?

Answer: Scientists know that increasing amounts of carbon dioxide in the atmosphere make the Earth warm up. If the amount of carbon dioxide keeps increasing, the Earth will eventually end up boiling hot with no plant or animal life at all, like Venus. The problem is that carbon dioxide can get into the atmosphere in many ways, from plants and rocks as well as from burning fuels. Scientists aren't quite sure yet how global warming will affect the amount of carbon dioxide coming from plants and rocks! So it's very hard to tell if global warming will slow down and stop, or just keep getting worse.

Question: I've heard that global warming is making ice at the Poles melt, and that will make global warming get faster. Why?

Answer: Light-coloured materials reflect much more heat and light than dark-coloured materials. The ice is light-coloured so it reflects lots of the Sun's energy back into space. When the ice melts, the sunlight shines onto dark-coloured ocean instead. The ocean absorbs more of the Sun's energy, making the Earth warm up even more, so even more ice melts.

Question: I read that we can stop global warming by planting trees. Is that true?

Answer: Trees absorb carbon dioxide from the atmosphere, making global warming slow down. But we must be careful, because some studies have shown that clearing the ground to plant trees can release more carbon dioxide from the soil than the young trees absorb. Other studies have shown that old trees absorb more carbon dioxide than young trees, so protecting old forests is better than planting new ones.

Question: My friend told me that global warming will make us colder and the Earth will end up like a giant snowball. How can that happen?

Answer: Your friend is getting two different ideas a bit confused. At the moment Britain is kept warm by the Gulf Stream, a current of warm water flowing north to us from the Gulf of Mexico. Scientists predict that global warming will disrupt ocean currents and stop the Gulf Stream flowing. Then Britain would have weather like Moscow, with lots of snow through the winter.

Scientists know that there have been times in the past when most of the Earth has been covered by ice – they call this a 'snowball Earth', but this happened when the continents were in different places. It happened when most of the continents were near the Equator; reflecting lots of sunlight back into space and making the Earth cool down. Scientists say it couldn't happen now because dark-coloured oceans near the Equator absorb too much energy from the Sun for the Earth to freeze over.

Planting trees slows down global warming, but only if carbon dioxide is not released from the soil. Protecting old trees is more effective.

- How does increasing the amount of carbon dioxide in the atmosphere affect the temperature of the Earth?

- Compare the amount of the Sun's energy reflected by ice and by the ocean.

- Why might global warming make Britain colder?

- Some scientists suggest that warmer temperatures may increase the amount of moisture in the atmosphere, increasing snowfall over the Antarctic and causing the area of ice to expand again. What effect would this have on global warming?

Activities

- Some studies have shown that plants grow more slowly, and absorb less carbon dioxide when it is warmer and there is more atmospheric moisture. Discuss how global warming may affect plant growth, and how this in turn would affect global warming.

- Role-play an *Any Questions* type TV programme where members of the audience ask a panel of experts questions about global warming.

Flashing the plastic!

The magnets you use at school are probably big heavy things made of steel, but magnets don't have to be like that. They can be made from plastic, rubber or ceramic so long as they have enough magnetic material in them. And they can be made so tiny that many people carry hundreds of thousands of them around every time they go shopping! This science magazine explains what is going on.

WHAT ARE CREDIT CARDS?

Credit cards are small plastic cards that you shop with, instead of money. They have a black strip on the back with a special number that a computer uses to find out lots of information about you. When you 'pay' with a credit card, a special reader in the shop reads the number from the card, then the shop charges the credit card company for what you bought, and the credit card company sends you a bill later.

WHAT THE BLACK STRIP DOES

The black strip on the back of a credit card is full of millions of tiny particles of iron, each one like a bar magnet less than one thousandth of a millimetre long. Before the card is given to you it is 'wiped' past a 'writer', with an electromagnet in it. The current in the electromagnet is changed quickly to make all the tiny 'magnets' in your credit card line up in a way that is special to your card – no other card will have exactly the same pattern of tiny 'magnets'. Then whenever you use your card, it is 'swiped' past a reader, which reads the special pattern of magnets on your card.

This card won't work any more. The black strip has been so badly scratched that the tiny magnets have been damaged and the reader can no longer read the pattern.

CREDIT CARDS AND SHOPLIFTING DEVICES

One of the most common ways to stop shoplifting is to attach small, paper labels, with magnetic tags in them, to the goods in the shop. The tags have a special pattern of magnetism in them, which makes a detector at the shop door ring. Pads at the checkout use a strong magnet to change the magnetism in the tag, so the pattern is destroyed and the alarm won't ring. You don't always see the pads because sometimes they are in the same machine as the bar code reader. But you might see a notice that says 'Do not put credit cards down on here.' If you do, the strong magnet will rearrange all the tiny magnets in the black strip on your credit card, and your credit card will be useless.

KEEP YOUR CARD SAFE AT ALL TIMES!

Criminals can make 'readers' that copy and store the numbers from the black strips on credit cards. If they copy your card, they can make a false card and use it to buy things using your money. So it is a good idea not to let your credit card out of your sight. Don't let a shopkeeper take it 'out the back' to check it, for example. Now banks are trying to find safer ways to shop than credit cards, because they are worried that criminals can steal credit card numbers from people doing internet shopping.

'SMART' CARDS

Soon all credit cards will be replaced with 'smart' cards – often called 'Chip 'n' PIN' cards. 'Chip 'n' PIN' cards still store lots of information about you, but they don't use magnets. The information is stored in a tiny computer chip, so it is much harder for thieves to copy information from the card, and steal your money. The computer stores a special four-digit security number that you can type in, instead of having to sign your name. It is really important to remember the four-digit number, instead of writing it down anywhere – that way thieves can't use your card because they can't forge the four-digit number.

- How big are the 'magnets' in the black strip on a credit card?

- Describe how the 'writer' puts information onto your credit card before you get it.

- What is under the pads that say 'Do not put credit cards down on here'? Why does it wreck a credit card?

- Give one advantage 'smart' cards have compared with ordinary credit cards.

- Shops lose billions of pounds a year because goods are stolen by shoplifters, and credit card theft costs millions of pounds too. Discuss what we could, or should, do about these things.

- You may listen to a 'walkman' with audio cassettes in it sometimes, or you may even record your own cassettes. See if you can find out how magnets are used to store songs on an audio cassette.

Hidden treasure

Setting the scene

Have you ever built your own electromagnet? You probably know how useful they are in scrap yards and in electric bells, but did you know that archaeologists like those in *Time Team* use electromagnets, and they have even been used to find buried treasure? Read this student's diary to find out more.

25TH JULY

I'm spending a week on an archaeology dig in the north of Scotland – I've always liked holidays that are a bit different! But I'm beginning to wonder if it was a mistake! It's freezing cold, there's a howling gale, it's raining and we are going to be living in tents all week! We are going to be investigating lots of bumps all over the grass, that the local historians think are probably Iron Age.

26TH JULY

The weather isn't much better, but at least it's not raining. I thought archaeology was all about digging things up, but we haven't done any digging at all! We spent all day surveying the bumps. Dr Spindler, who is in charge, says you can find out a lot about an area without digging it up. Some people were drawing maps to show the positions of all the bumps. I was using a magnetometer to make a map of the magnetism in the soil. The magnetometer looks like a thick broom handle with a heavy box the size of a lunch box on top, and I spent all afternoon walking up and down with it, touching the end of the 'broom stick' onto the soil at 1 metre intervals! Tomorrow I'm going to ask how it works! Now I'm tired!

27TH JULY

My birthday! Sunny all day! I asked how the magnetometer worked, and I wish I hadn't – it's very complicated. Apparently there's an electromagnet in the magnetometer, but the core is a tube of water. Tiny magnets in the water line up with the magnetic field of the electromagnet. Then the electromagnet is switched off, and the tiny magnets in the water line up with magnetic fields in the soil. As the tiny magnets move they make a current in the electromagnet wire. A computer measures the current to 'picture' the magnetic field in the soil. Dr Spindler had to explain it three times before I understood it! Apparently things like ditches, walls and metal objects have different magnetism from the soil around them, so they show up as black marks on the computer 'picture'.

28TH JULY

We did some digging today, in the places where things showed up on my magnetometer survey. It's much more exciting than I thought it would be! The round bit turned out to be a stone wall from an Iron Age house, and the blob in the middle was ash from a fireplace. The Iron Age people used to build a round stone wall about a metre high, then make a tall sloping roof of reeds and grass, with a hole to let out the smoke from the fire. We're going to have a go at building a small Iron Age house tomorrow.

30TH JULY

We're going home tomorrow. I'm a bit sad really. It's been interesting. The best bit was when I found the metal brooch. Before it was cleaned, it just looked like a lump of rusty coloured mud. It's exciting to think that I found it and I was the first person to touch it for more than 2000 years!

Qs

- What is the core of the electromagnet in a magnetometer made from?

- Give three things that would show up in a magnetometer survey, because they have a different magnetism from the soil around them.

- Walls, ditches and areas of ash only give small readings on a magnetometer. What size reading do you think an iron brooch would give? Can you suggest a reason why?

Activities

- Soils or rocks containing magnetic iron oxide will give higher magnetometer readings than the surrounding soil. Discuss what else an archaeologist might look for to help him or her decide if variations in magnetism are natural or man-made?

- 'It is very important to find out about how people lived in the past.' Discuss why you agree or disagree with this statement.

The Northern Lights

If you have made an electromagnet, and looked inside an electric motor, you may well have guessed that electricity and magnetism are always linked together. Sometimes, if you are really lucky, you may see a dramatic demonstration of this link between electricity and magnetism, in the sky. Read this holiday diary to find out more.

February 17th

Well, here we are in Finland! I still wish we had gone to Spain, but the hotel is nice and I think it might even be fun. There's snow everywhere, and Mum says we can go skiing and driving snowmobiles.

February 18th

Last night was amazing. Mum woke me up at 2 a.m. and said 'Just come outside.' We watched the Northern Lights, and I can't describe them. There were these flickering, glowing lights all over the sky to the north. They started red and orange, with streaks of blue and mauve and yellow, then after about ten minutes it faded to a greenish yellow glow over the sky that gradually faded away. It was amazing! I found out that there's a guy in the hotel who's come here just to see the Northern Lights.

February 19th

The guy who's come to see the Northern Lights told me today that the scientific name for the Northern Lights is 'Aurora Borealis', and you can see them near the South Pole as well, only there they're called the 'Aurora Australis', or Southern Lights. He's called Kris, by the way. Apparently, as well as heat and light, the Sun sends out streams of electrically charged particles called the solar wind. When the electricity of the solar wind gets to the Earth's atmosphere it is attracted down to the poles of the Earth's magnetic field. That's why you only see the Northern Lights near the North Pole. The electrical particles make gases in the Earth's atmosphere glow, and give out coloured lights.

February 20th

I've been wondering about the Northern Lights again. I asked Kris why they don't last all the time if they are caused by a solar wind from the Sun, and why we didn't see any at all last night. He said the solar wind keeps changing, as the Sun sends out different amounts of electrical particles. It's all to do with sunspots and solar flares. When there are lots of sunspots the Sun sends out lots more solar flares, and then we see the Northern Lights more often. Each solar flare only lasts about half an hour or less, so the Northern Lights don't last long either.

February 21st

We spent all afternoon on snowmobiles. It was magic, but my arms ache. We saw the Northern Lights again last night, and it was just as good as on Monday. I keep thinking of more things to ask Kris about them. He told me today why there are so many different colours. It turns out that there are lots of different types of gases in the atmosphere, at different heights above the Earth. The colour of the Northern Lights changes when the electrical particles hit different types of gas, and that depends on how strong the solar wind is and how far through the atmosphere the electrical particles get! Complicated, isn't it! It's great to watch, though.

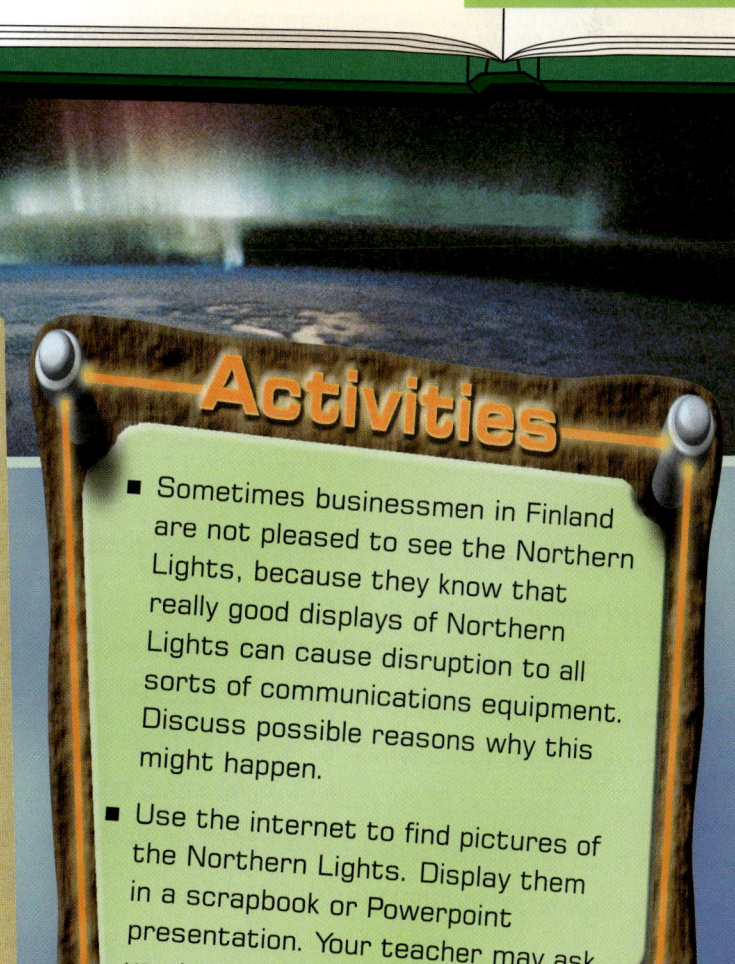

Qs

- What is the scientific name for the Northern Lights?

- What causes the Northern Lights? Why can you only see them near the Earth's magnetic poles?

- How are the Northern Lights linked to sunspots and solar flares?

- Explain in your own words why there are lots of different colours in the Northern Lights.

- The strength of the Earth's magnetic field changes gradually over time. How would you expect this to affect the Northern Lights?

Activities

- Sometimes businessmen in Finland are not pleased to see the Northern Lights, because they know that really good displays of Northern Lights can cause disruption to all sorts of communications equipment. Discuss possible reasons why this might happen.

- Use the internet to find pictures of the Northern Lights. Display them in a scrapbook or Powerpoint presentation. Your teacher may ask you to write some information to go with your images.

Newton's eye

We see objects because light reflects from them and enters our eyes. We can change what we see in lots of ways, such as using glasses to improve poor vision, or using microscopes to see very tiny things. But some people try to change what we see by changing the shape of our eyes. Read these imaginary diaries to find out more.

Setting the scene

I am still trying to find out about how I see things. I want to know if it is my eye or my brain or both that are important. I have dissected sheep's eyes and they have a lens at the front. I think the back of the eyeball acts a bit like a screen to show a picture of what comes through the lens. I looked hard at the Sun, then looked at a screen and found I could still see a 'picture' of the Sun. **(Editor's note: NEVER try this at home, folks!)** I think my brain remembers what my eye saw. I won't do that experiment again though because I couldn't see anything for three days afterwards and I was afraid I was going to stay blind.

I have tried pushing blunt needles and blunt metal plates into my eye socket and pushing on different parts of my eyeball with them. **(Editor's note: DON'T try this either!)** I have found that when I push hard on my eyeball I see coloured spots and rings. I am still wondering why this happens. I think it is because the pressure in my eyeball is changing, because the spots and rings are fainter when I don't push so hard, and bigger if I push with my finger instead of a needle.

Extract from Isaac Newton's diary, 1665

SAFETY!

NEVER copy any of Newton's experiments with his eyes – he was very lucky not to blind himself. In Newton's time many scientists did dangerous experiments on themselves and many scientists injured themselves, sometimes permanently.

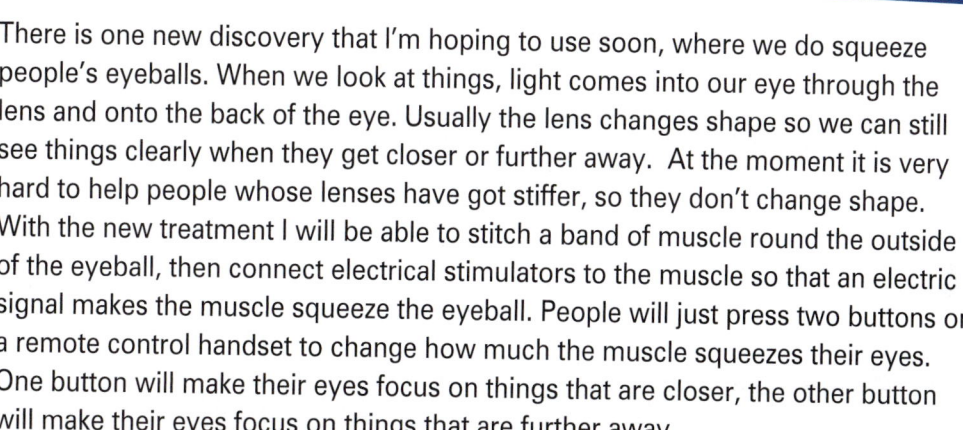

I have been reading about Newton – and I still use some of his discoveries today! The coloured rings or spots or patterns patients see often help me work out what diseases they have. We don't put needles behind people's eyes though!

There is one new discovery that I'm hoping to use soon, where we do squeeze people's eyeballs. When we look at things, light comes into our eye through the lens and onto the back of the eye. Usually the lens changes shape so we can still see things clearly when they get closer or further away. At the moment it is very hard to help people whose lenses have got stiffer, so they don't change shape. With the new treatment I will be able to stitch a band of muscle round the outside of the eyeball, then connect electrical stimulators to the muscle so that an electric signal makes the muscle squeeze the eyeball. People will just press two buttons on a remote control handset to change how much the muscle squeezes their eyes. One button will make their eyes focus on things that are closer, the other button will make their eyes focus on things that are further away.

Jaspreet Pang, eye doctor, 2003

Qs

- How did Isaac Newton find out what different parts were in the eye?

- Why did Newton think he could still see a 'picture' of the Sun even after he looked away? What do you know about looking at the Sun?

- How does Jaspreet Pang use Newton's discovery about coloured rings and spots?

- What people can be helped by the new treatment that squeezes the eyeball?

Activities

- Whenever doctors or scientists invent a new treatment they don't know whether or not it will work properly. Discuss who should try out the new treatment first, who should decide whether someone is offered the treatment, and what we should do if it goes wrong.

- A friend has just telephoned you saying they don't know whether to have the new electrical treatment or not. Work in pairs to role-play the telephone conversation you will have with your friend.

Amazing lasing!

Has your teacher used a special type of light called a laser to show you how light travels in straight lines? Lasers are used for many things much more exciting and useful than just laser pens or laser pointers. These people who work with lasers tell us more.

Setting the scene

Interviewer Jeff, you work for a car manufacturer, and you use lasers every day, don't you?

Jeff That's right. I use a cutting laser to cut out some of the really fiddly metal parts. A computer controls the laser beam so it cuts in the right place.

Interviewer You're telling us a laser can cut metal? I thought it was just a type of light.

Jeff It's a very special sort of light with only one colour in it. The light doesn't spread out as much as white light, so all the light energy can be focused on a tiny area. It makes the metal so hot it melts.

Interviewer That sounds really dangerous. Is it safe to have lasers in schools?

Jeff The ones in schools are nothing like as bright and powerful as the ones I use. They have a lot less energy, so they are safe to use, as long as you don't shine the beam into anyone's eyes.

Interviewer I see. Thank you, Jeff.

Interviewer Good afternoon. You use lasers to remove tattoos?

Dr Solanki That's right. About half the people who have tattoos wish later that they hadn't. I help them get rid of them.

Interviewer How does the laser help?

Dr Solanki The laser light passes through skin, but is absorbed by the ink in the tattoo. The energy breaks up the ink into really tiny particles, which are carried away by the body's immune system.

Interviewer Does it hurt?

Dr Solanki A little. Some people say it feels like tiny drops of hot fat on the skin.

Interviewer And does the tattoo disappear completely?

Dr Solanki It depends on the tattoo. After all, tattoos were meant to be permanent. Most tattoos leave a faint mark that is easy to cover up with make-up.

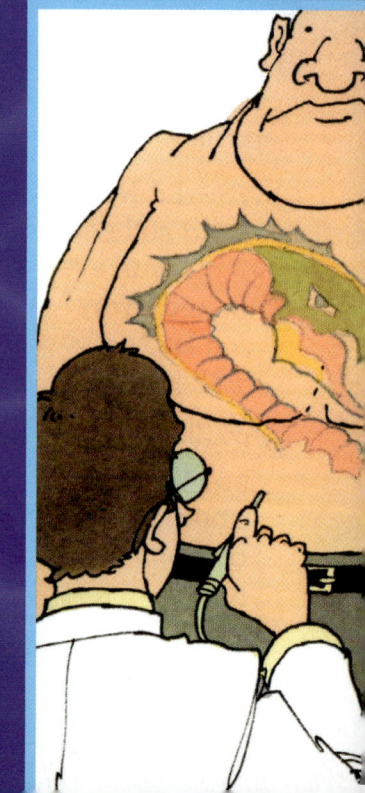

Interviewer Jade, I've seen you at work, and I found it fascinating. Just explain to our listeners what you do.

Jade I use something called laser stereolithography to make plastic models from a clear plastic liquid.

Interviewer I can't even pronounce that. Tell us what it is.

Jade Its other name is 3D layering. I make a computer model of what I want to build, split into lots of thin layers, a fraction of a millimetre thick. The laser 'paints' the first layer on the top of the liquid plastic, and the plastic sets where the laser shines on it. Then the model is lowered a fraction of a millimetre and the laser 'paints' the next layer, and so on, until I have a complete model.

Interviewer Is that a good way to make models?

Jade Well, it's expensive and it's slow. A complete model can take 12 hours to build. But designers find it very useful. It's cheaper to make one model, to test how good the design is, using 3D layering rather than ordinary machines, even though ordinary machines are better for making thousands of copies once the design works.

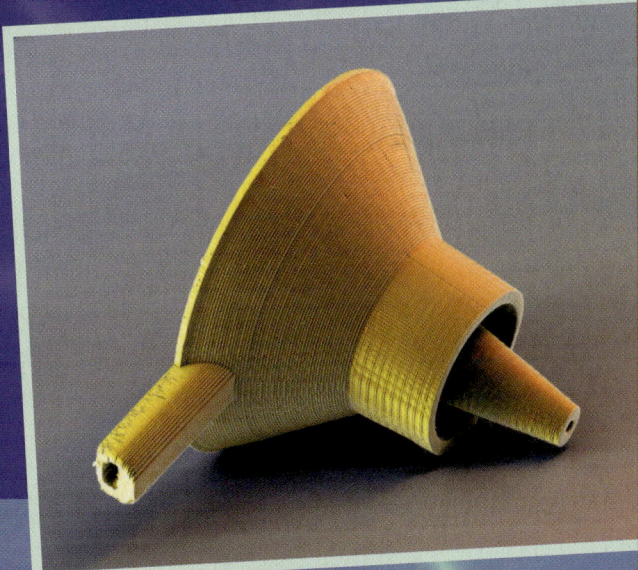

Interviewer Well, I'll let you get back to your models!

Qs

- List three things that lasers can be used for.

- Give one difference between laser light and 'ordinary' light.

- Black tattoo ink absorbs all colours of laser light. Coloured inks only absorb some colours of laser light. Which do you think is harder to remove, a black tattoo or a coloured one? Give a reason.

Activities

- Discuss the safety precautions Jeff, Dr Solanki and Jade might take when using their lasers.

- Lasers are also used in lots of other places, such as speed cameras to catch speeding motorists, laser eye surgery, or CD players. Do some research to find out more about one of these uses of lasers, or about any other use you choose.

They do it with mirrors

Setting the scene

We first began to build instruments to help us see better and further in the 1600s. One of the latest and most exciting instruments is the Hubble Space Telescope. So what makes the Hubble telescope so special? Our interviewer talks to a scientist from the Hubble Space Program.

Interview with Jake Lyman, 24th April 1990

Interviewer Well, Jake, the Hubble Space Telescope is due to be launched tomorrow. How do you feel?

Jake Really excited. We have been waiting a long time for this day.

Interviewer What makes this telescope so special?

Jake It will be the first telescope in space. We're hoping for the best pictures anyone has ever seen.

Interviewer This may sound like a stupid question, but wouldn't it be a lot cheaper and easier to build a telescope on the ground?

Jake It's a very sensible question. And, yes, it would be cheaper and easier. But we know we will get clearer pictures from space. When we look at the stars from Earth, clouds get in the way and, even when the sky is clear, the stars seem to twinkle. That's because dust, water vapour and moving air currents distort the light from the stars, so the pictures come out fuzzy. We won't have any problems like that from a telescope in space, above the Earth's atmosphere.

Interviewer Thank you Jake, I shall talk to you again when we get the first pictures from the Hubble Telescope.

Interview with Jake Lyman, 30th April 1990

Interviewer Well, Hubble is safely launched. Are you pleased with the pictures it is sending back?

Jake No, they're awful. I can't tell you how disappointed we all are. They're much worse than the pictures from good ground-based telescopes.

Interviewer So having a telescope in space doesn't work then?

Jake It's not that. It's nothing to do with being in space.

Interviewer What is it then?

Jake Hubble has a huge mirror that collects light from stars and channels it down to the scientific instruments at the centre of the telescope. The mirror is supposed to be perfectly smooth, but it wasn't ground properly. The fault in the mirror is less than one fiftieth the thickness of a human hair, but it's enough to make the pictures fuzzy.

Interviewer So will Hubble be scrapped now?

Jake No, of course not. We will just have to solve the problem of the mirror.

Interviewer How will you do that?

Jake I don't know yet, but we'll find a way.

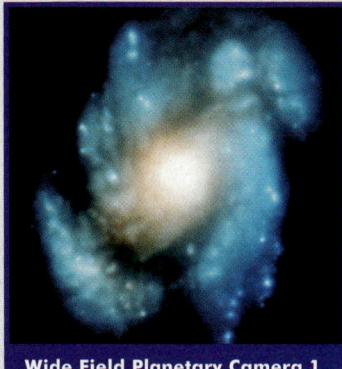

Wide Field Planetary Camera 1

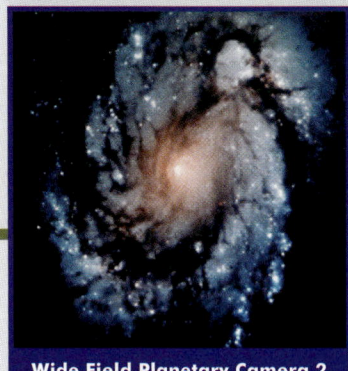

Wide Field Planetary Camera 2

Interview with Jake Lyman, December 1993

Interviewer Hello, Jake. It seems a long time since we spoke. How's Hubble's pictures?

Jake They're amazing – even better than we had hoped for.

Interviewer So how did you solve the problem with the mirror?

Jake We fitted a 'contact lens'. We designed a set of tiny mirrors, that sat in the way of the beam of light from the main mirror, and changed its direction slightly, so it was going where it would have been if the main mirror had been right in the first place. We have called the 'contact lens' COSTAR, short for Corrective Optics Space Telescope Axial Replacement.

Interviewer So how long will Hubble last now?

Jake I don't know. Ten or fifteen years at least, I should think.

- Why are pictures from a telescope in space clearer than pictures from a telescope on the ground?

- Explain in your own words how the Hubble Space Telescope works.

- Why were the first pictures from the Hubble Space Telescope fuzzy?

- How did scientists solve the problem, to get clear pictures?

Activities

- The Hubble Space Telescope repairs were carried out by astronauts doing 'spacewalks'. Discuss the dangers, excitements and alternatives to doing repairs in this way.

- Use the internet to find out some more about the Hubble Space Telescope or about some of the space programs it has been used for.

Smashing sounds – fact or fiction?

Sounds are made by objects vibrating. You know lots about describing different sounds, about how sound travels through different things, and you have probably investigated how people hear sounds. You may even have heard stories about loud sounds breaking things. Today we ask an expert whether these stories can possibly be true.

Sound Experts

The walls of Jericho

There is a story in the Bible that tells about trumpets being blown outside the walls of the city of Jericho and then the walls just falling down. I am often asked if this is a true story. The simple answer is we just don't know. We can't even do experiments to find out because we don't know exactly how tall the walls were, or what they were made from, or exactly what sound the trumpets made. But I do know that sounds can be used to break things. People have shattered glasses by playing, or singing, exactly the right note next to them.

So how does that work?

Sound makes things vibrate, and different pitch notes make things vibrate at different speeds. All things have a special speed that they vibrate well at; it's a bit like pushing a swing, if you push at exactly the right speed (the right time between pushes) the swing swings more and more, but if you push at the wrong speed the swing stops. If the sound played next to the glass is exactly the right note, the glass vibrates more and more until it shakes itself to pieces!

This is loud enough to make you go deaf, if you were near enough without ear defenders.

Fatal sounds

You know that loud sounds are dangerous because they can damage hearing, but extremely loud sounds can kill! There are many tales of people dying because they climbed up to the bells in large churches or cathedrals while the bells were ringing. At least some of these tales are true. We think the people die because the sounds make blood vessels inside their brains vibrate so much they burst. We certainly know that people who survive very loud sounds, such as very big guns going off right next to them, often bleed from the ears.

Shaking to bits

Kidney stones are hard 'stones' formed in the kidneys from crystals of minerals and proteins in the urine. They are very painful. The only way to remove them used to be an operation, but now they can be treated without surgery. Doctors send a narrow beam of very high pitched sound, called ultrasound, at the kidney stones. The sound passes through the soft tissue of the body without damaging it, but it makes the hard kidney stone vibrate so much that it shatters into small pieces. The small pieces can pass out of the body with the urine.

Squeaky clean

Can you imagine using sound to clean something? It's a very good way to clean a piece of machinery that is too delicate or too difficult to take apart. Very high pitched sound, called ultrasound, is used to make the machinery vibrate, to shake the dirt off; it's a bit like shaking a rug, but easier – and it works better!

- Give two examples of things that can be broken by sounds.

- You can be killed by standing in the same small room as ringing church bells. Suggest a reason why the same bells don't kill you if you are a long way away.

- Explain why a sound has to be the right note to shatter a glass. Use the model of a swing being pushed in your answer.

- Do you think the pitch of the sound used to treat kidney stones matters? Give a reason.

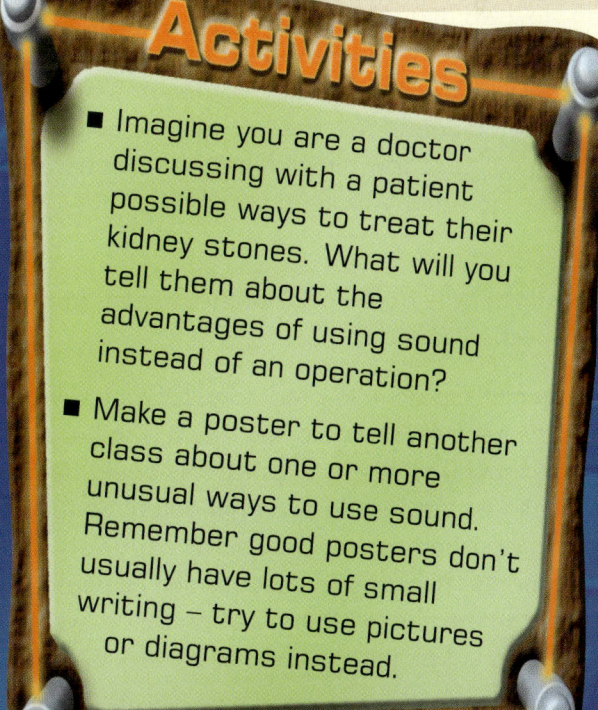

Activities

- Imagine you are a doctor discussing with a patient possible ways to treat their kidney stones. What will you tell them about the advantages of using sound instead of an operation?

- Make a poster to tell another class about one or more unusual ways to use sound. Remember good posters don't usually have lots of small writing – try to use pictures or diagrams instead.

Did you hear that?

Objects vibrating produce sounds. You know from experience that some sounds 'carry' through the air much better than others, but have you ever thought how this affects our lives, or the lives of other animals? Read these questions and answers from the pages of a science magazine to find out more.

Question: I walked to my local fair recently and I noticed that I could hear the deep notes of the fairground music long before I could hear the screams of the people on the rides. Why is this?

Answer: The screams are high-pitched sounds that make the air particles vibrate rapidly. These quick vibrations don't travel very well because they are easily disturbed by other movements of the air, such as the wind blowing. The slow vibrations of deeper sounds are much less likely to be disturbed by other air movements. All sounds that are meant to travel a very long way, such as whales calling to each other, or jungle tom-tom drums or Australian didgeridoos, are low-pitched sounds. The crocodile's mating call is so low pitched that we can't hear it, but we can see the water vibrating!

Question: Why do all baby animals seem to make high-pitched sounds?

Answer: It's to do with predators. Often baby animals need to 'shout' to say 'Mum, I'm here, I'm hungry', but if all the local predators could hear them as well they would soon get eaten. So they use high-pitched sounds, that don't carry very far. Their mother hears them because she remembers roughly where she left them, but predators won't hear them unless they happen to be very close.

MUM!

Question: The government wants to build a new, longer runway at my local airport. They say it will be quieter. How can a longer runway be quieter?

Answer: Aircraft have to speed up to 'lift-off speed' before they can leave the ground. On a short runway the pilot has to rev the engine hard to get the aeroplane going fast enough before the end of the runway, and it makes a lot of noise. On a longer runway he does not have to rev the engine so much and it makes less noise.

Question: I watched a television programme about the Loch Ness monster the other day where they used sonar to try to find it. How does sonar work? Also they said their sonar was better than old sonar because the frequency was higher. What did they mean?

Answer: Sonar stands for 'SOund Navigation And Ranging' which really just means finding your way round using sound. The scientists sent sound waves down to the bottom of the loch and timed how long it took the waves to reflect back. The deeper the water, the longer the sound takes to come back. The scientists used the sound waves to make a map of the bumps on the bottom of the loch. Bumps that move around aren't rocks – but they might be the Loch Ness monster.

Higher-frequency sonar gives a sharper, less blurred 'picture' of the bumps on the loch bottom, but it does not travel so well through the water, so it is not as good as lower frequency sonar in very deep water.

Qs

- Which sounds would you expect to hear first, as you walked towards a noisy school disco?

- Why do baby birds make high-pitched 'cheeps' instead of low-pitched sounds?

- Explain why low-pitched sounds travel further through air than high-pitched sounds.

- Give one advantage and one disadvantage of using higher-frequency sounds for underwater sonar surveys.

Activities

- Imagine that an airport near your home wanted to make its runways longer. Discuss how it might affect you. Decide whether you would be for or against the expansion.

- Making runways longer is not the only way to reduce the noise from airports. Think about how sound travels through, and reflects from, different materials, and then role-play a meeting between airport authorities and local residents to decide on ways of protecting a nearby village from the noise of an airport.

Cracking good alarms

You know that sounds are made by objects vibrating, and you have learned many ways to change sounds, by changing the vibrations causing the sounds. Usually we think the vibrations are useful because they produce the sounds we want. Sometimes, though, it is the sounds that are useful because they tell us about the vibrations. Scientists are discovering new ways to use sounds like this, as early warning systems.

Coal miners' early warning system

Miners have known about sounds as early warning systems for hundreds of years. All pit props (the supporting posts and beams used to stop the sides of the tunnel collapsing) used to be made of wood, but in the Industrial Revolution metal-working skills improved and it became possible to make metal pit props instead, that were stronger and lasted longer. However most miners didn't want them. They said that the wooden pit props were safer because if part of a tunnel was about to collapse the wooden pit props would creak and groan first, giving miners time to run out of that part of the mine. Metal pit props did not break so often, but when they did there was no warning, so miners were more likely to be trapped or injured.

Predicting earthquakes

Scientists find it almost impossible to predict earthquakes, because most of the time the Earth just seems to move without any warning. However, recently scientists have discovered that sometimes the ground creaks before it moves. It is not a creak that you can hear, it is a very low-pitched sound – too low pitched for human ears to hear, but it can be detected by scientific instruments. Scientists are doing research to collect more information. Perhaps one day they will be able to build an earthquake warning system.

Creaking knees!

Many people say that their knees, knuckles or other joints creak or crack when they move. Some people even make their knuckles crack on purpose, for fun! Some doctors say that most people creak or crack at some point in their lives and it's nothing to worry about, others say that healthy joints are supposed to move smoothly and quietly and creaking and cracking is a sign that joints are not moving as smoothly as they should.

A team of Canadian scientists from Toronto, doing research to listen to creaking knees, have made some interesting discoveries. They have found that all knees, healthy and injured, creak whenever they bend, but the creaks are usually too low frequency for us to hear. The scientists recorded lots of knees creaking and put the creaks into a computer that changed them to audible sounds. They found that the creaks from healthy knees and injured knees sounded quite different. Knee injuries are very common in sports like cricket, cycling and football, where athletes put a lot of strain on their knees. The Canadians say their research can check for signs of knee damage before an injury happens, and can say whether an injured knee has healed enough for an athlete to return to training without the injury recurring.

Knee injuries are very common in sports like football and cricket.

Qs

- From the article, list three ways in which sounds can be used to give warning.

- Give one advantage of metal pit props and one advantage of wooden pit props.

- What type of sounds do all bending knees make?

- Suggest some research you would do, if you were working to develop a way of predicting earthquakes.

Activities

- Knee injuries are common in many sports, but proper training and 'warming up' makes them less likely. Discuss what you do to reduce your risk of injury in the sports you play.

- Do a survey of the pupils in your class, to find out how many of them have had creaking or cracking joints at any time. Does it happen all the time, only occasionally or only after an injury?

Acknowledgements

The author and publishers would like to thank the following for
providing photographs.

Cambridge University Library: *68;*
Corbis: Michael Cole *40,* Ralph White *54;*
Corel 427 (NT): *10;* Corel 638 (NT): *10;* Corel 752 (NT): *31;*
   Corel 434 (NT): *44, 45;* Corel 449 (NT): *44;* Corel 39 (NT): *47;*
   Corel 412 (NT): *48;* Corel 11 (NT): *66;* Corel 604 (NT): *77;*
Digital Vision WT (NT): Jeremy Woodhouse *25;* Digital Vision 5
   (NT): *26;* Digital Vision 2 (NT): *27;* Digital Vision 9 (NT): *37;*
   Digital Vision WW (NT): Jim Reed *67;* Digital Vision 7 (NT): *67;*
   Digital Vision 6 (NT): *72;* Digital Vision 6 (NT): *75;* Digital
   Vision 12 (NT): *79;*
Flexon: *41;*
Getty Images: *45;*
Gordon Baer/Cincinnati/USA: *16;*
Lawrence Berkeley National Laboratory: *35;*
Martyn Chillmaid: *9, 63;*
NASA: *73;*
Photodisc 67 (NT): *25;*
Science Photolibrary: Garry Watson *17,* David Scharf *19,*
   Sheila Terry *50,* Dr Ken Mcdonald *55,* Eye of Science *71;*
TopFoto: C M Dixon/HIP *65;*
USGS: *52*

Metal ores photograph page 39 by John Taylor.

Picture research by John Bailey